A HANDBOOK FOR INVENTORS

CALVIN D. MacCRACKEN

A HANDBOOK FOR
INVENTORS

*How to Protect, Patent,
Finance, Develop, Manufacture,
and Market Your Ideas*

CHARLES SCRIBNER'S SONS
NEW YORK

To my father,
Henry Noble MacCracken,
and my mentors,
Ted Edison and Tony Nerad,
who taught me the fun of ideas.

Copyright © 1983 Calvin D. MacCracken

Library of Congress Cataloging in Publication Data

MacCracken, Calvin D., 1919–
  A handbook for inventors.

  Bibliography: p.
  Includes index.
  1. Inventions—Handbooks, manuals, etc.
  2. Patents—Handbooks, manuals, etc. I. Title.
  T339.M26 1983   608′.068   82-42668
  ISBN 0-684-17906-7

This book published simultaneously in the United States of America
and in Canada—Copyright under the Berne Convention.

All rights reserved. No part of this book may be reproduced in any
form without the permission of Charles Scribner's Sons.

1 3 5 7 9 11 13 15 17 19 F/C 20 18 16 14 12 10 8 6 4 2

Printed in the United States of America.

Contents

Introduction ix

1. **Protecting Your Idea Yourself** 1
   Your Diary 2
   Sketches 4
   The Preliminary Disclosure 8
   Making a Rough Model 10
   Trying It Out 13
   Evaluating the Idea 13

2. **Getting Your Patent** 17
   The Five Types of Patents 22
   The Prior Art Search 23
   The Specification and Drawings 25
   The Claims 44
   Filing 52
   Interference 54
   Infringement 54

3. **Financing—The Part You Can Do** 56
   The Sting 57
   Finding Investors 61

The Business Plan   64
Forming a Corporation   67
Other Sources of Money   68
Starting Up   69
Getting a Partner   72
A Patent Exchange   73

4. Financing—The Big Leagues   76
Venture Capital   77
The Partner   79
Corporate Finance   84
The R&D Partnership Tax Shelter   85
Going Public   86

5. Development on Your Own   90
The Development Process   92
Designing for Production   94
Laboratory Testing   97
Field Testing   98
Cost Estimating   100
Certification and Approvals   101

6. Contracting Development and Manufacturing   104
Product Development   105
The Development Contract   108
Communication   112
Investment in Special Tools   113
Industrial Design   115
Subcontracting the Manufacturing   115
The Subcontracting Agreement   119

7. Licensing or Sale   124
Finding a Licensee   124
Should You License, or Sell?   127
Negotiating the License Agreement   128
The Option   132
How Will Your License Do?   133

8. Setting Up Your Own Plant 135
   Personnel   136
   Motivating Workers   137
   Equipment   138
   Suppliers   139
   The Bread-and-Butter Line   140
   Customer Payments   141
   Money Management   141
   Financial Difficulties   143

9. Marketing by an Outside Firm 146
   Original Equipment Manufacturers   146
   Exclusive National Distributors   149
   Trademarks   153
   The Importance of Marketing   154

10. Marketing It Yourself 155
   Direct Sales   156
   Pricing   157
   Wholesale/Retail   158
   Reps   161

11. New Fields to Conquer 164
   A Few Case Histories   167
   The Failures and the Successes   172

Appendices 175
   Appendix A. Some Important Patents Issued in the United States, Primarily to Independent Inventors, in the Last 100 Years   175
   Appendix B. Patent Searchers   181
   Appendix C. Groups That Help Evaluate Ideas   182
   Appendix D. Product Development Companies   184
   Appendix E. Some Companies Looking for Inventions   186

Appendix F. Sample License and Option Agreements 188
Appendix G. The Support Network 197
Appendix H. Recommended Reading 201
Appendix I. Inventors' Shows 209

Index 210

Introduction

You are probably reading this book because you have been inspired by an idea and want to do something with it. Thomas Edison said, "Invention is 2 percent inspiration and 98 percent perspiration," and it is to cut down on all that sweat and frustration that this book was written.

In Edison's day it was easier, in some respects, to make an invention and get a patent. It took him only 45 minutes to sit down and write out a patent in longhand, draw a sketch, and mail it off to the U.S. Patent Office.

But other things were more difficult for him. When I worked with Theodore Edison, Thomas's son, at Calibron Products, his research and development laboratory in West Orange, New Jersey, he told me of his father's hard, hard work, his continual disappointments, his lack of money, his deafness, and his desperate searches for (as well as love of) materials. Late in life, when he was to be honored at a grand dinner for his achievements in the use of electricity, Edison was asked what he would like as a gift. He quickly replied, "A cubic foot of copper." He got it—all 559 pounds of it—and today it sits in the Edison Laboratory Museum in West Orange. Materials, he knew, were the key to invention.

Today, with Alexander Graham Bell's invention, you can reach tens of thousands of suppliers of every imaginable material and component, eager to help you locate at no charge just what you need to implement your idea. So while filing a patent may be more complicated now, the opportunities to invent things and make money on them have increased if you know what to do. Because of the availability of materials, the tree of invention has grown, with thousands of new branches and twigs, and each material is the seed of a new tree, so there is now a virtual forest of opportunities.

Take EPDM (ethylene propylene diene monomer), a synthetic rubber-plastic elastomer material developed about ten years ago. It costs more than common synthetic rubber, but it has unique properties of resistance to high temperatures and atmospheric degradation. Dozens of inventions, patents, and new products have grown from EPDM, including membrane roofing that lasts virtually forever, solar collector/absorbers rolled out on reels, window gaskets that will keep out wind and rain, automobile bumpers that absorb crash impact, heater and radiator hoses that will not burst and leave you stranded in your car.

In fact, EPDM is now compounded in hundreds of varieties, with benefits to a wide circle of experts, consultants, manufacturers, extruders, engineers, marketers, installers, and customers. Had you known of EPDM ten years ago, who knows what you could have done with it, or, for that matter, what you might still do with it.

As a professional inventor for thirty-seven years (thirty-five of them as president of my own company), I have developed over 275 new products, both for myself and others. After starting with Edison in 1940, I moved to the General Electric Research Laboratory during the war years, where I worked on then-secret jet engines for our nation's defense. I also headed General Electric's Creative Engineering Program, where I selected promising young inventors from their engineering ranks and trained them in the classroom. They were given inventive engineering assignments and apprenticed to successful inventors within the company on a three-month rotating basis.

The qualities I looked for in these young men (there were no women working in the engineering departments of General Electric at that time) were creativity as part of the family background, developed hobbies, love of games, originality of thinking and expression, a gift for mechanical visualization, and the ability to solve simple puzzles.

I learned that it isn't the ability to invent that counts so much as the will to do so. Many talented people who could invent are seduced by the security of a corporate job and channel their thinking into narrow product lines and conservative policy.

It takes a strong ego and personal commitment to be a free-lance inventor. You must be willing to fail repeatedly to risk scorn and bankruptcy. The positive side includes the joy of creativity, a quickening of the heartbeat, an exhilaration better than that offered by any stimulant. When one of my original inventions works for the first time, I have a feeling closely akin to falling in love, a desire to repeat over and over to myself, "I've done it, I've done it, I've done it!"

Sometimes I think of my eighty patented and over eighty unpatented inventions as if each were my own child. But that's the wrong way to think, because whereas a child demands an unqualified commitment, an invention must be dropped if, for any of many reasons, it's just not going to work out. If you must make a decision to drop what you're doing, remember that it's not your creativity that has been rejected, and you will soon come up with something better. The commitment is to creativity, not to every form it may take.

You probably have an idea that excites you, which is great. But you are understandably concerned about diverting time and money from your other commitments, what others will think of you, and what is really involved in making your idea succeed. Most of all, you want to know whether it will sell and how to go about achieving sales with a reasonable profit/effort ratio.

The most important decision you'll have to make concerns which aspects of the work you propose to handle yourself. There are five key operations in getting your invention to the money-making stage. These are patenting, fundraising, development, manufacturing, and marketing. You can handle all five, or you can contract any or all of them out for others to do. If you are not going to handle either the manufacturing or the marketing, you will probably have to license or sell your invention. To enable you to make your decisions on these issues and to follow them through, I describe them in detail.

What you decide to do depends very much on the product and your own capabilities and aspirations. I have always chosen to do my own product development, at least to a certain stage, because without working on the development of my ideas I would not know my product, would overlook possible improvements, and would miss out on most of the excitement. As for the other operations, I have worked out many different combinations for what I would do and what would be farmed out, depending on the situation.

This book gives you the information you will need in these areas; it will show you how to do it yourself, where to go to find others to do it for you, and how to work with them. It points out the pitfalls, the shortcuts, and the ways to save money. My only purpose is to stimulate your creativity, show you how to market that good idea you have, and help you find the exhilaration, achievement, and financial rewards that can come with invention. Perhaps someday the United States will honor its inventors as the British do (Sir Robert Watson-Watt, radar; Sir Frank Whittle, jet engines). Until then inventors will be happy if they know they've increased employment and can afford to go on to the next invention.

CHAPTER 1

PROTECTING YOUR IDEA YOURSELF

As the writer of this book, I am protected against potential copiers because my publisher obtained a copyright for me from the federal government in a very fast, simple, and inexpensive procedure of which most authors are hardly aware.

As an inventor, I regard getting a patent as all-important, for a patent is the essence of an invention. In fact, an unpatented device is rarely considered an invention, even though the dictionary defines "inventing" as "originating, as a product of one's own contrivance." "Patenting" is not part of the definition; but without this step, it's tough—a patent is the only thing that can protect your work and ensure that it will be recognized as yours.

Protecting your idea by patenting it is not a simple task. It involves words and, usually, pictures, and your idea may not be easy to put into satisfactory words and pictures. An idea or invention evolves as you work on it, and you must be able to document its evolution to prove you were the originator. Then, if no one else has thought of your idea before and if the ordinary person "skilled in the art" would not have thought of it, you are entitled to protection.

YOUR DIARY

The best and most convincing way to document your invention is to keep a witnessed diary. This diary, a bound notebook, should be a complete record of everything having to do with your idea: its background, its conception, its development, the setbacks as well as the advances, your calculations, your plans, with whom you talked about it and what they contributed, drawings, sketches, test results, photographs, what you've read or seen that relates to it, and anything else you think of. A school notebook with lined paper and a red line at the left margin is often used, the date of entry being put in the margin.

Don't worry about keeping your diary clean and neat, putting your scribbles on scratch paper and then transposing them neatly later—if you remember. Your record will be far more convincing if you keep a real working journal, going straight to your diary to put in the results of your work and using it to do your calculations as to how big something should be or how much it will cost. Just don't keep any scratch paper lying around, so that any time you pick up your pen or pencil—and it should be often—you will use your diary.

Leave room at the bottom of each page for your signature and the date, even though you have dates in the margin. You may have several days' entries on one page. Then, periodically, take your diary to someone you can trust and have him or her sign the page as well, after the words "Read and understood by me" along with the current date.

If you keep this up until you have completed the idea's development, you will have the next best protection to filing a patent application. Sending registered letters, having a notarized disclosure, getting witnesses to sign a document, or even telling your attorney about the invention are helpful only in establishing a date and the fact that you were

TITLE 115 F Heat Storage Eutectic Salt Tank
Project No. AR-1
Book No. 303

From Page No. 18

4-30-80 There is a great deal of noise during and after nucleation of the salt as it starts to freeze while being cooled down. Salt is forming around the impeller of our stirring pump which we use to prevent stratification of the hypo salt. The amperes of the pump motor rises in less than one minute from about 1.0 to 1.4–1.6 as crystals on the impeller produce drag. Also crystals on the pump housing hit the impeller causing noise and vibration.

5-2-80
A.M. We have made changes in the pump. Brian suggested putting in an acrylic star-shaped baffle with nylon bushing to steady the shaft and prevent vibration. I suggested using a rubber impeller instead of metal. We also used our old heater wire but increased to several turns directly in the impeller area. The use of rubber I have done in small pumps before to prevent rubbing noise. Brian has made these parts and reassembled them in the salt tank. We hope this combination will stop crystal growth, vibration and noise in the impeller area.

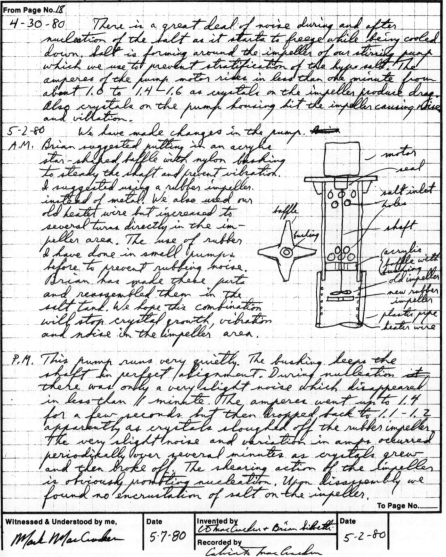

motor
seal
salt inlet holes
baffle
bushing
shaft
acrylic baffle with bushing
old impeller
new rubber impeller
plastic pipe
heater wire

P.M. This pump runs very quietly. The bushing keeps the shaft in perfect alignment. During nucleation there was only a very slight noise which disappeared in less than 1 minute. The amperes went up to 1.4 for a few seconds but then dropped back to 1.1–1.2 apparently as crystals sloughed off the rubber impeller. The very slight noise and variation in amps occurred periodically over several minutes as crystals grew and then broke off. The shearing action of the impeller is obviously promoting nucleation. Upon disassembly we found no encrustation of salt on the impeller.

Witnessed & Understood by me, Mark MacCracken **Date** 5-7-80
Invented by CB MacCracken + Brian Silvetti **Recorded by** Calvin MacCracken **Date** 5-2-80

A page from an actual laboratory diary showing work done on a device to pump salts used for thermal storage. Note the use of entry dates and sketches, the witnessing, and the signatures of the inventor and recorder.

involved. These measures do not document the step-by-step development of the invention or idea—which you must be able to demonstrate in order to prove that you were truly the inventor—nor will they show, for example, that the real date of your invention (which usually falls some time before you write up your disclosure) was earlier than that of some other inventor working on the same idea.

SKETCHES

No matter how poor you are at making sketches, put them in your diary. They are better protection than any signature because they are distinctly yours and they tell several stories.

First, sketches make the idea clearer to you and to any reader. They are the best way to communicate the story you are trying to tell.

Second, they help you improve your idea. Lots of good ideas are lost because they aren't unique enough by themselves to be patentable; yet if they were combined with something else that is also unpatentable, the combination might be patentable. For example, the combination of the lead pencil and eraser was an important patent, although each alone was not new. Making a sketch of a whole pencil, when all you were thinking of was how to put the lead and wood together in a better way, might have helped you think of the combination.

Third, drawings are required in most patents except those involving drugs or specific materials, and even there drawings of a process are often used. Since you will have to do them some time or have them done (in which case they might not be just what you had in mind), you should start with them early.

Fourth, drawings point out the places where more work is required. They lead you to your next step.

Finally, drawings are believable evidence to a jury or judge trying to determine what your contribution to the invention was. You cannot hire someone to do the inventing for you and then claim to be the inventor.

Sketching improves greatly with practice. Most inventors are poor sketchers: there are very few Leonardo da Vincis. Never be embarrassed to try, and be sure to do them all in your notebook diary even if you later draw a line through them.

There are three kinds of sketches: the schematic diagram, the cross-sectional engineering drawing, and the perspective sketch.

Schematic Diagrams

The schematic diagram is used primarily for ideas pertaining to electrical or fluid systems, for organization charts, and to show sequences of operations or anything else in which one element of a system must be explained in relation to another.

The accompanying drawing shows, in schematic form, a system for using ice storage to shift some of the electrical air conditioning load in an office building from daytime to nighttime. If used widely enough, this would eliminate the need for new generating plants, nuclear or coal-burning, and save everyone money. Here the system is the key to understanding how the invention works. A schematic diagram must show the relationship and scope of the parts plus the various modes of operation.

Engineering Drawings

The cross-sectional engineering or architectural drawing shows front, side, and top views. Architects call these drawings the *elevation* and *plan* views of a structural element or component. For a patent application, these

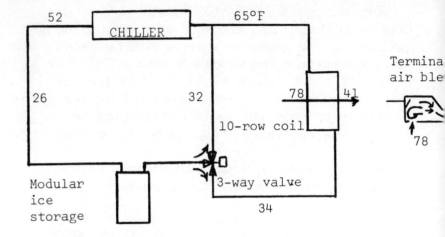

Schematic piping diagram of an air-conditioning system using ice storage for lower costs.

would not include dimensions but rather give identifying numbers for all the parts to be referenced in the required verbal description (detailed in Chapter 2). Such drawings are needed to make clear the structure of a component, what it consists of, and what it does.

The drawing here, of the ice-storage component of the system depicted in the previous sketch, shows how it is made of low-cost materials; how it stores cooling and gives it up; how it can freeze ice solid without breaking, stressing, or compressing anything; and how it melts the ice. The cross-sectioned drawing shows, theoretically and two-dimensionally, how the component would look if cut in half. Materials, supports, fasteners, relative locations and thicknesses, and many other features are denoted and described.

Perspective Sketches

The perspective sketch is provided when a three-dimensional representation is needed for understanding. For

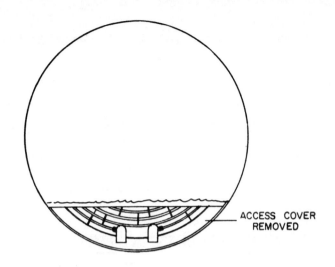
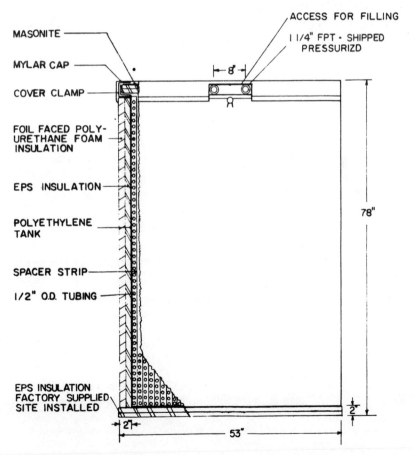

Cross-sectional engineering drawing with legends explaining the elements of the ice tank/heat exchanger.

example, the drawing below depicts a mechanical room showing the modular ice tanks, the chiller, the ducts, and the piping system in such a way that the relative sizes and arrangement can be judged.

You should consider these three types of sketches and use in your notebook diary those that can be helpful in understanding and communicating your concept.

THE PRELIMINARY DISCLOSURE

Besides the regular entries and the sketches, you also need a written summation or disclosure of your idea in your diary. This disclosure, like the patent application described in Chapter 2, has several parts: background, purpose, general description, detailed description, and alternatives.

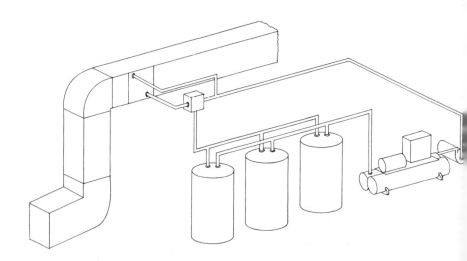

Perspective sketch shows the piping, ducting, and components of the system arranged in a room.

The *background* is the brief history of the art or related subject matter, as you know it, leading up to your idea. The *purpose* is an explanation of what your idea achieves and why it is valuable. The *general description* is an overall view of how your idea achieves its goal, with reference to the three types of sketches. The *detailed description* makes full use of characters or numbers on the drawings, denoting the various parts, and gives a step-by-step account of how everything shown interacts as a system, component, structure, or method. The *alternatives* discuss ways other than the one you described in detail in which your idea might be achieved.

This preliminary disclosure in your diary is important for all the reasons the sketches are; in addition, it may be the only complete description of your idea ever written. Not all patentable ideas are patented. In fact, I would estimate that only a small percentage are. There are many reasons for this: the negative ones include lack of money, lack of market interest, technical or cost problems, or the inventor's own loss of interest; among the positive reasons could be secrecy or the need to make a head start.

Remember that the patent protection provided under the U.S. Constitution is designed to give the inventor a seventeen-year monopoly in return for a full and complete disclosure. If, however, you have a process or material you can keep secret even though your idea is being marketed, you may be better off just to keep it to yourself. The formula for Coca-Cola was never patented for this reason—it could be kept secret because the taste came from a concentrate mixed up infrequently, in a special way, at an unknown location. But these cases are still relatively rare. Most chemical companies, for example, patent their processes and additives, knowing that sophisticated analytical equipment can unravel many of their secrets. The best bet in most cases is to disclose everything and get the protection you are entitled to. If you don't, one risk is that someone else will come along and patent the same thing, which may squeeze you out.

When a product has a brief life, a good head start is all you need and a patent is not worthwhile. Design patents on toys, dolls, fad items, or this year's model of clothing rarely make sense. In the case of Rubik's Cube, for example, a well-established trademark is far more valuable than a patent on how the parts interlock so they will rotate in two directions.

In other cases, fear of a patent's being held invalid in a legal contest argues against filing. The new patent law adopted in December 1980 provides for a much less expensive determination of invalidity for a contested patent within the U.S. Patent Office. Because the threat of costly litigation will be reduced, more inventors may be encouraged to get their patents.

MAKING A ROUGH MODEL

Making a model or prototype of your idea is an essential step and goes right along with your diary. Many people say, "Oh, I don't know anything about making things," yet they can cook, change a tire, sew on a button, light a fire, replace a switch, paint a chair, make a sign, wrap a package, assemble a bookcase, and so on. You can generally do whatever you want to do—not perfectly, but well enough. "Where there's a will, there's a way!"

There are many shortcuts to making models. Think of the available materials and components that you can buy. There are suppliers and manufacturers out there who have what you want if you can just find them. Use the Yellow Pages, the fourteen-volume *Thomas Register*, regional manufacturing directories, directory issues of trade magazines, arts and crafts catalogs, and do-it-yourself catalogs in every field; also check with hardware and electronics stores. Talk to your friends who have access to purchasing agents. Go down to the local high school and ask if you can talk to the vocational or shop teachers. Don't be afraid to reveal necessary parts of your idea so as to get the help you need. Your idea is protected by your diary, provided that you have kept it up diligently.

At General Electric, when I ran the Creative Engineering Program, I noticed great differences in openness among corporate inventors. It seemed that those who were most open and willing to seek advice and help were those who were most successful. Therefore I resolved to be open and take the risk.

You'd be surprised what you can do with cardboard and Scotch tape. Art stores sell Bristol board, which takes paint well, is stiff, and yet can be cut and bent neatly. Vinyl sheeting can be cemented better than other plastics. Auto and hardware stores carry vinyl adhesive. Vinyl also comes in rigid sheets, tubes, and pipes. Fasteners of all types, such as screws, bolts, quarter turn fasteners, and snaps, are available. Screws come in many varieties, including machine, sheet-metal, self-tapping (they make their own thread), Allen head, and Phillips' head. They are available in different lengths and thicknesses. It may seem complicated, but free catalogs with all the data are available from manufacturers. Find a place to store your catalogs; they will accumulate rapidly. Electrical parts and tools are now much easier for do-it-yourselfers to handle. Radio, TV, and hi-fi stores have new wiring materials, components, and tools that save time and look professional.

Prototypes sometimes work for years, serving the intended purpose, but they may not be developed further and marketed for any of a number of reasons. For example, when my plant offices were burglarized for the fifteenth time by crooks who knew how to elude the usual alarms, my chief engineer made an interesting observation. Noting that it was the typewriters that were always stolen, he suggested connecting 12 volts through the 110-volt typewriter motor to hold in an alarm relay but not run the motor. Then, when the cord was pulled out of the 12-volt wall socket, the alarm would go off. (To use the machine, the secretary would change the plug position by placing it in a specially rigged outlet.) The engineer built his setup from standard parts, and the next three burglary attempts were foiled as a result. If the reader wants to build a business on this idea, it's okay with me. Many people have their own ingenious though not quite marketable devices working around their homes or businesses.

Try to use standard parts whenever possible. If you need a special part, a skilled sheet-metal worker or cabinetmaker can often construct it fairly inexpensively if you explain what it is you are trying to do. If you wire up a panel yourself and then take it to an electrician who does wiring for a living, he or she might redo it for you for a small charge. I've had these sorts of things done for me many times, though I've always tried to staff my own model shop with skilled people.

Another way to obtain a model is to hire someone at a model or machine shop to make it for you or assign that person a share of the profits in return for the service. Like anything else, your success depends upon whom you select. Hiring the job out can work, but such arrangements should be kept to a minimum and payment should be, at least in part, on a profit-sharing basis. Take the time to make specific drawings of what you want, including dimensions. Most model shops reject divided design responsibility because they cannot then give you a fixed quote of the cost of making your model. They have learned from experience that inventors seldom realize how much a model costs, and they want this settled right at the beginning.

TRYING IT OUT

Testing is something you must do yourself or at least supervise personally. You will never know your product unless you have experienced its testing and development, its troubles and successes. If someone else comes up with the solution to a major problem, he or she is likely to become your coinventor, with rights equal to yours.

Ideas and inventions are so varied that they can be tried out in many places other than a commercial laboratory. Most are first tested in basement workshops. They may start out in the form of a hobby, taking a minimal investment of time and money until results begin to look promising. Working evenings and weekends can make the process a slow one, particularly when tools and space are inadequate. There is the danger that the inventor's enthusiasm will fade before enough can be done. Conversely, this may be the time when you develop vital improvements for your invention.

It is true, however, that small company laboratories have the best record of success when it comes to the tougher problems. They can be found in the Yellow Pages under the listing "Laboratories—Research and Development." Again, profit sharing may be a key to getting good outside help. A company that works in the general field of the idea can be more efficient than a laboratory that takes on all fields.

Full development—usually a long, hard job—will be thoroughly covered in Chapter 5. The first tryout is fun and exciting and must be made a part of your diary.

EVALUATING THE IDEA

Since you're reading this book and, I hope, have resolved to keep a diary of your idea, you are already part of a select few among inventors. Robert Onanian, editor and publisher of *Invention Management* and founder of the Institute for Invention and Innovation, says, "Only 1% of inventions ever make it to the market. If we could somehow bring it up to 2%, we'd have increased new jobs and the economy enough to solve all of our nation's financial problems." Of course, we're all interested in being among that 1 percent (and possibly 2 percent), but the question is: How *good* is your idea? As Onanian puts it, "Most of the inventions that make it are just better ideas than the others."

Does that mean that there's no use working on it and improving it—that it will either make it or it won't? Not at all! Few successful writers had their first piece accepted. Few of the businesses that are started succeed the first time. What is needed is evaluation. If your idea is not good enough, you can do any of three things: drop it and come up with a better one, improve it by making changes and additions, or learn how to convince people that it *is* good enough. All three strategies can work. First, though, you must know where you actually stand. Talk to the people who really know. The editors of trade publications are an excellent source of information. Find out what magazines, trade papers, newsletters, and professional journals there are in your related fields through the directory (periodical index) carried by your local library. Subscribe to these publications; call up their editors; go see them. Tell them what you want to do and get their advice.

Probably the best bet is the trade show, exhibit, convention, or whatever it is called in your field. Here you can talk to buyers and sellers, service organizations and parts suppliers, new and established, large and small, innovative and traditional companies. Find out how large the market is for your type of product or service. Determine what manufacturers are biggest and the size of others in the field. Learn how the marketing and selling are done. The people at

Protecting Your Idea Yourself

these shows are intent on selling their own products and not likely to be a threat to you, particularly if you reveal only what is needed to get their evaluation and stress that it's all very "iffy." For example, you could ask a salesperson, "If I had a device that would do _____for $_____, would you be able to sell it?" Then ask, "What features are most important? What other features are needed? Could you sell a deluxe model at twice the price? How would it be serviced?"

Pick out a booth exhibiting a similar or related product or component and ask to talk to an engineer. Tell him you're working on a product to do a certain thing. Don't say "inventing," because that word arouses legal concern and suspicion. Ask what technical matters he'd be worried about, and had he ever heard of anyone developing, writing about, making, or selling such a product. Remember that for a product to be successful the market must find the product, not the product create a market. All you are doing is trying to let the market know that you and your invention exist. If the market doesn't respond at least in a small way once that's done, there may be a real problem.

With these and other evaluations, you should decide whether to drop your idea, improve it, or plow ahead with it as it is. For the best chance of success, try to improve it. One thing I've noticed repeatedly is that if an invention is to be successful, there must be more than one reason for its existence. An old inventor friend once came to me with a slit metal tube that slid over a milk carton top, keeping it airtight so the milk would stay fresh longer, keeping it from spilling if you carried it with you, and making it easier to carry because of the tube's tight grip. He had three reasons already, but the invention still wasn't patentable. However, by magnetizing the tube so it would stick to the refrigerator door when not in use, and adapting it to grip heavy cards and invitations as a reminder, he not only enabled it to qualify for a design patent but also made it much more salable.

On the other hand, don't put in more innovation than you need. There's a fine line between being different enough to be patentable and too complex to be marketable. I invented a home heating system called Jet-Heet that had both a unique, jet-type oil burner and a small, flexible, high-velocity duct-heating system. Problems with servicing the oil burner damaged the market for the duct system, which had great appeal initially. Even after thirty years, few dealers will handle it, though the name has been changed and a less efficient, conventional burner is used. Had I simplified it in the first place, I'd have made so much money I might not have done a lot of other fascinating things.

CHAPTER 2

GETTING YOUR PATENT

Like an author on publication day, you as an inventor will never be happier than when you learn your patent is granted and see it in print. Each year, over 40,000 books are published in this country and 65,000 patents awarded to U.S. inventors, but because more than half the patents are from foreign countries, creativity in verbal and technical fields is roughly comparable. Countless manuscripts and inventions never see the light of day, either because they lack sufficient novelty or quality or because their creators are not tenacious enough.

It is important that you learn as much about patents as everyone knows about books. Many newspapers have a whole section on books every Sunday, plus daily reviews. Patents, if you're lucky, get one short column a week by one business reviewer who does not even begin to evaluate them. There are no Nobel Prizes or other major awards for inventors; the few prizes there are go mainly to large companies. Rube Goldberg gave inventors a reputation as nuts, and media coverage often reinforces this. Yet we inventors know better, and the general public does too.

When an inventor has a patent, he really has something important.

Unfortunately, the federal government has decided to make patents more costly to get and to keep by increasing patent fees. The average filing fee (before you know whether or not you will get a patent) went from $85 to $300 on October 1, 1982—a 253 percent increase. The issue fee (after you know you will get a patent and what it will say) went from $145 to $500—a 245 percent increase. Maintenance fees of $400 after four years, $800 after eight years, and $1,200 after twelve years have been added, much like the annual fees of most foreign patents. Although independent inventors, small businesses, and nonprofit concerns must pay only half these fees, it's still a large increase.

These are small costs, however, compared with what a good attorney will charge to prepare a patent application, and you are throwing your money away in most cases if you use inadequate legal help. I used to pay $500 to $1,000 for the preparation and filing of my patent applications twenty or thirty years ago. A few years ago this cost began to exceed $5,000, so I worked out an arrangement with my attorney whereby I write the whole thing, including claims, and he edits or rewrites it as necessary. This has cut my cost to something between $1,000 and $1,500. (Of course, costs of this kind can vary widely.)

Therefore, if independent inventors (including those in very small companies) are to continue to live and innovation is to thrive, they must learn to prepare their own patent applications; in addition, the Patent Office must not only liberalize its formal rules but also instruct its examiners to offer active help to applicants seeking the protection to which they are entitled. I am lobbying with the Patent Office and Congress on this now, and I hope you will all assist me by writing your representatives and the Commissioner of Patents.

Getting Your Patent 19

Some patent examiners will give help in a personal conference. Often an inventor wants to patent the structure of a novel device, but the examiner finds in his search that the parts and their arrangement are similar to something done before and therefore the claim cannot be allowed; if certain changes were made in the claim, however, it could be allowed. A former competitor of mine in the solar energy business was once seeking to improve upon my flexible roll-out rubber collector/absorber—he wanted to make it easy to remove the webbing connecting the tubes so that the clamps could be applied to the ends of the tubes and the mat turned. If he called these weakened webs "tear strips," a structural term, he would be referring to the entire length of the mat, but by using the functional adjective "separable," as the examiner suggested, he could refer to only the clamping sections and rewrite the claim to describe the method of attachment rather than the structure of the mat. Examiners can give this sort of help and a great deal more with virtually no additional work on their part. The examiner then becomes more of a consultant or collaborator than an adversary, which is very much in accord with the national interest in increased innovation.

Preparing your own patent for review by an attorney is the best way to handle the procedure in most cases, so you must discipline yourself for the work this requires. This approach will weed out those who wouldn't spend the time to follow their ideas through to market anyway, and it will push you into getting your development, manufacturing, and market planning done.

Always remember that if you have sold your product or published an article or advertisement about your idea more than one year before your filing date, you are not entitled to valid protection in the United States and many other countries.

Learn what you need to know about writing patent

United States Patent Office

Des. 219,896
Patented Feb. 9, 1971

219,896
COMMERCIAL TOASTER FOR RESTAURANTS
Manfred Hegeman, Nyack, N.Y., assignor to Calmac Manufacturing Corporation, Englewood, N.J.

Filed Jan. 12, 1970, Ser. No. 20,866

Term of patent 14 years

Int. Cl. D7—04

U.S. Cl. D81—10

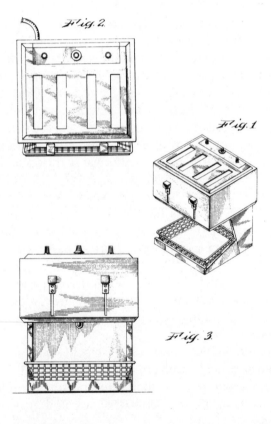

Design patent of a drop-through commercial toaster.

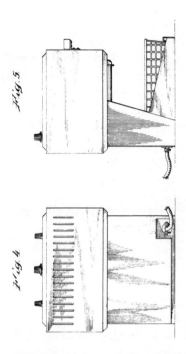

Fig. 5

Fig. 4

FIG. 1 is a perspective view of the commercial toaster for restaurants as seen looking downwardly toward the front and right side;
FIG. 2 is a top view;
FIG. 3 is a front elevational view;
FIG. 4 is a rear elevational view; and
FIG. 5 is an elevational view of the left side.
I claim:
The ornamental design for a commercial toaster for restaurants, substantially as shown and described.

References Cited
UNITED STATES PATENTS

D. 176,117	11/1955	Hanson	D81—10
D. 190,783	6/1961	Heilweil	D81—19
D. 191,024	8/1961	Buday	D81—10
D. 200,517	3/1965	Kratz	D81—10

OTHER REFERENCES

Institutions, June 1959, p. 117, Toastmaster automatic 4-slice toaster No. 1D2.

J. PAUL GUERTIN, Primary Examiner

applications from this book and others listed in Appendix H ("Recommended Reading") at the back of the book, and by reading lots of patents, particularly those in your field. These can be ordered from the Patent Office after you have copied down the patent numbers from the name plates and literature available at trade shows. But most important of all, sit down and try to write the application yourself. Then consult an attorney; the names and addresses of attorneys and agents registered to practice before the U.S. Patent Office are available in a pamphlet from the U.S. Government Printing Office (see page 204).

THE FIVE TYPES OF PATENTS

There are five types of patents: (1) design, (2) structure, and (3) process/method/system patents, which are the most common; (4) patents on materials or combinations of materials; and, recently, (5) patents on living cells or combinations of cells.

A design patent is relatively simple—principally a drawing and protection of the general appearance and visual arrangement of the invention. Once, inspired by my success with a hot dog cooker in which the frankfurters rotated slowly in the valleys between heated rollers, I developed a commercial toaster called the Toastbasket. Instead of popping the toast up, the Toastbasket let it drop into a heated basket when it was done; then more slices could be toasted and the basket could be used for carrying. By hiring an industrial product designer, I developed an attractive package for the invention. The patent protection for this was a design patent because the visual appearance of the toast falling into a portable basket was the novelty, not the toaster structure or toasting method. The accompanying illustrations show the simplicity of the patent.

Getting Your Patent 23

The structure patent is often called a mechanical patent. In addition to drawings, there is an expanded explanation of how the idea works, starting with a brief abstract, the history behind its invention, its advantages, identification of the drawings, a description of the invention with detailed reference to the drawings using characters or numbers, examples of how it works, alternative ideas or uses, and finally the legally worded claims that describe the protection you have been granted.

Process/method/system patents involve the way a group of components and/or materials work together. They usually include schematic diagrams showing electrical wiring circuits, piping arrangements, flow charts, mixing schedules, a sequence of processes, or a combination of these.

Combination-of-materials patents are common in the drug field and usually have no drawings, just descriptions of the materials and how they are mixed and used.

Living-cell patents used to be limited to asexually created plants but now cover a variety of new developments, such as gene splicing, in biochemistry and related fields.

THE PRIOR ART SEARCH

"Prior art" refers to previous inventions in your field. If your idea is inventive (that is, not obvious to someone skilled in that art), is new, and someone hasn't invented, published, or used it in public before, it can be patented. This is what you pay the U.S. Patent Office to find out when you file your patent application.

However, it is often considered wise to make your own search first, or to have it made by a professional searcher, with the object of saving yourself time and money. Although you can do it yourself, especially if you live near Washington and have the time to learn the procedure, it is probably best at first to have it done by a searcher. A list of searchers is included in Appendix B. After describing your invention in words and drawings, you can tell the searcher to limit the investigation cost to $100 or $200 and to advise you whether and why he thinks it would be wise to spend more.

Now, I have never had a search made before filing; in eighty U.S. Patents issued, I've never had reason to regret the decision. Only one of my applications has been abandoned. There are many reasons why this may work for some and perhaps not for others. First, being active commercially in related fields, reading the trade press, and asking others in a position to know of any such products, I have a pretty good idea of what is available. Of course, my idea could have been invented and abandoned, but my knowledge of the field tells me my idea is too good for that.

Second, my search would not necessarily be the same as the examiner's search, so I can't rely on it. And third, I'd rather spend my time and money on something positive (preparing the patent application) than something negative (looking at what others have done that is close to my work). An inventor is a person who takes risks, is prepared for many failures, and finds ways around whatever problems arise. I'd rather go forward until I hit the problem the examiner hands me, and see if I can't overcome it, than worry about trying to find some problem that probably doesn't exist. Besides, this invention is going to succeed only if I'm excited about it, not if I'm being conservative. In fact, I've erred much more on the side of not asking for enough patent protection than asking for too much. A well-written specification can sway an examiner more than the old references he finds.

Your experience may not be the same as mine, however, especially if you are not as well informed as you might be about developments in your field. So for your first few inventions it would probably be wise to have a prior art search made. As you gain more knowledge and confidence, you may find, as I do, that the money and energy required for the search are better spent in other ways.

THE SPECIFICATION AND DRAWINGS

If you've kept up your diary, in some ways you can write your patent application better yourself than a patent attorney could. This is your creation; it's your patent. It's not what someone else thinks it ought to be. It should be written as you see it.

The Abstract

The abstract is a very brief, one-paragraph description of the idea. Pretend you are making a transatlantic person-to-person call on a weekday to tell a friend what it is and why you invented it; then maybe you'll be brief enough. It's mainly there to let the reader decide whether or not to look at the rest of the application.

As a help to both of us, I am going to write a patent application on one of my new ideas. I've never heard of anyone else proposing it and neither have experts in the field with whom I have discussed it. Even if the basic idea is old, I have some new wrinkles that should give me the protection I need. The application I am outlining can easily be tailored to both a structure patent and a process/method/system patent, as we shall discuss in the section on claims (page 44).

ABSTRACT

This invention relates to storing the winter's cold, or part of it, for use in air conditioning during warm weather. A flat-bottomed, pondlike excavation is bulldozed out of level ground; a plastic, waterproof liner is laid down; and a grid of plastic tubes is laid out over the liner and connected to inlet and outlet headers. A layer of the excavated dirt is bulldozed back over the tube grid; another layer of tubes is laid down parallel to the lowest grid; alternate layers of tubes and dirt are applied up to a plane beneath the original ground level; a layer of insulation is put down; a waterproof membrane to shed rainwater is applied; a final backfill of dirt is put on; the inlet and outlet headers are connected to piping, pumping, and ambient air–heat exchange means; the system is filled with water by soaking hoses to within about 12 inches of the insulation; and the pump and ambient air–heat exchanger is operated whenever the ambient air is below the freezing point of water. Alternatively, dual tubes may be driven into the ground vertically at spaced intervals, with antifreeze liquid going down one tube and up the other. The ground, acting as a large, insulated ice-storage bin, will preserve most of the ice until warm weather, when cooling may be extracted from the ground by pumping the antifreeze mixture from the ground tubes to the chilled-water system of the building that is to be air conditioned.

This provides almost free air conditioning, reduces the peak loads of electrical utilities, allows for matching the cooling demand of the building by controlling the rate of withdrawal from storage, and provides for use of the required ground space for parking, lawns, driveways, etc. by preventing heaving through expansion space in the dry volume beneath the insulation. Where there is not enough cold weather to charge up the ice storage, it may be aided by the use of a small chiller of the type used in ice rinks. Even in places where almost no below-freezing weather is encountered, the chiller can be 20 percent of the size of a standard air-conditioning chiller and the operating cost about 20 percent of the standard.

Well, that was not so good! My transatlantic phone bill is way too high. I've actually put down much that should be saved until the next part of the specification; so here's a second try.

ABSTRACT
This invention relates to a method of storing winter's cold for use in air conditioning a building during warm weather. It depends on freezing a depth of wet ground by pumping an ambient air-cooled antifreeze solution through horizontal or vertical layers of plastic tubes buried under or driven into said wet ground. During warm weather, the ground-cooled antifreeze solution supplies cooling to the building's chilled-water system. Means are provided to prevent ground heaving, thereby allowing for other uses. Small supplemental mechanical cooling means may advantageously be provided in areas of insufficient cold weather.

That's about right. Once you've done a good abstract, chances are you'll do the rest. Notice the use of convenient Patent Office jargon: "said," referring to something you've mentioned before; "means," a broad word that doesn't tie you down when you're bringing up well-known technologies; "advantageously," meaning you'll discuss the advantages later. Always describe your idea as an invention.

The Background

Now it is time to discuss the background, the history of others who have worked in this field, the patent numbers of any related patents you may know, or any publications. *This is a must.* Include how you got the idea, what it means in its field or to groups of people as a whole, and other comments to indicate how this invention should be related to what has gone before. Much of this may be adapted from the disclosure in your diary. Here's my try.

BACKGROUND OF INVENTION

Before the advent of electric household refrigerators, natural block ice was cut on lakes or ponds and then stored, surrounded by sawdust for later delivery in ice wagons and use in food refrigeration. There was little loss by melting because of the natural self-insulation of the large block of ice and its high latent heat of fusion, 144 BTUs per pound.

Researchers in Quebec and Ottawa have been trying to devise methods to freeze large blocks of ice naturally inside buildings and then use their melt water for summer cooling. These methods were recently described at a conference at Argonne National Laboratory, Argonne, Illinois, in June 1981, as was a project at Princeton University in Princeton, New Jersey, which has received wide publicity. The latter project involves ski-resort-type snow-making machines that, during cold weather, manufacture snow into an excavated pit until it is mounded up. Known as an "ice pond," the snow is then covered with insulation and a portable structure is placed over it. Then, when cooling is needed, the melt water is pumped to building's chilled-water system, the return water sprayed over the snow pile, which cools it as it trickles through.

I have noticed in relation to the mechanically refrigerated ice rinks manufactured by my company under patent #3,751,935, 3893,507 and #3,636,725 that if the rinks are not properly insulated, ice may form in the ground under them over a period of years and become very difficult to remove. A method to prevent this buildup of ground frost, which causes heaving damage, is shown in my patent #3,910,059, in which I installed plastic tubes directly under the rink insulation and pumped in a lightly heated antifreeze solution.

This gives the background, the references, and how I came to be familiar with ice formed in wet ground. There is no requirement for brevity here and you should put in everything you know that is pertinent. If you expect the patent examiner to play fair with you, you should not cover up something similar that you know about. You can lose all your rights if it is discovered that you pretended ignorance intentionally.

The Purposes of the Invention

The next section of the application describes the purposes of your invention—its advantages, its possible uses, and its goals.

OBJECT OF THE INVENTION

The main purpose of this invention is to reduce the costs of air conditioning by storing winter cold for use in warm weather, when it can cool and dehumidify a building or group of buildings.

An equally important purpose is to save energy by using natural cooling instead of electrical energy generated from fossil or nuclear fuels.

A third purpose is to reduce the peak summer loads of electrical utilities, which must build new plants to meet peak demand. These peak-load generating plants are the least efficient by as much as 2 to 1 and tend to increase the use of imported oil.

Another purpose is to provide a cooling storage facility that is inexpensive to build, efficient to operate, and makes use of available space by freeing the area above for other purposes.

An additional purpose is to provide a long-term cooling storage facility whereby [another much used word], even though there is insufficient natural cooling or none at all, a small, inexpensive compressor may be run efficiently and at low electric rates for thousands of hours during the off-peak cooling periods to provide air conditioning during warm weather.

Another purpose is to provide cooling when needed at much higher rates than can be done conventionally and to modulate the output as required anywhere between 0 and 200 percent or more of design specifications.

A final purpose is to provide a means of determining what percent of the cooling storage reservoir is frozen and whether it is completely frozen.

The objects and purposes are now set forth. You should include all you can think of because this may serve to prove that you had uses in mind that your claims may not specifically cover, allowing a possible loophole. The judge at the time of a challenge to your patent wants to determine exactly what you were thinking of when inventing and may rule against your adversary by the "doctrine of equivalence" if he thinks you anticipated a particular use or structure. In other words, judges usually try to get at your real intention and may go beyond the limitations of words if they feel that the defendant was doing something equivalent.

Please note other convenient and persuasive buzzwords in the above paragraph: "object," "purpose," "claim," "use," "adversary," "structure," "anticipate," "equivalence." As in any field, knowledge of the jargon tends to show the writer's familiarity with the subject. Of course, overuse creates a negative impression, and the more experienced you become, the less you need to rely on such words.

Description of the Invention

In this section, start by putting in a brief overall description of your invention. This is the place for the material that was edited out of the abstract; therefore I won't repeat mine here.

Then a list setting forth the title and purpose of each drawing (or figure) is given, followed by the real meat of the whole specification, a detailed description referring to each figure and to each number designating a part of a figure.

DESCRIPTION OF DRAWINGS

Figure 1 is a cross-sectional drawing showing the horizontal-layer method of seasonal cooling storage in frozen ground.

Figure 2 is a cross-sectional drawing showing the vertical dual-tube method of the same thing as Figure 1.

Figure 3 is a schematic flow diagram showing the relationship between natural cooling and supplemental compressor cooling with the seasonal storage facility.

Figure 4 is a perspective view of the connections between ground loops and header manifolds.

When you do the drawings, start with pencil sketches, making sure you have plenty of eraser. I've picked a cross-sectional sketch as Figure 1. The Patent Office, even with its conservative rules, likes a jazzy drawing for No. 1, a sort of "show-stopper" that makes one wonder "What is that?" (Fortunately, the office now accepts informal drawings at the time of filing; these can be formalized when your

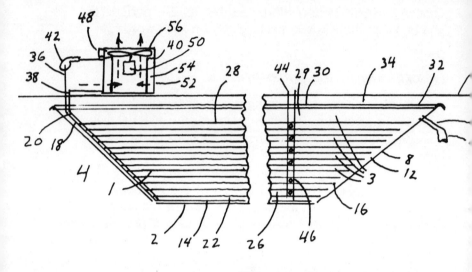

Figure 1.

Informal patent drawing of a winter ground-freezing system for seasonal air conditioning.

invention has been allowed. The sort of drawings reproduced here are perfectly acceptable for the application.)

Figure 1 shows a cross-sectional drawing of a pit 2 bulldozed out of the ground 4 from ground level 6. The endwalls 8 of the pit 2 are at low angle 10, for example, 1 foot vertically for 2 feet horizontally, to permit a bulldozer to conveniently negotiate the slope of the endwalls.

Don't worry about being repetitive, aim for complete clarity; it's not an essay. The numbers 2, 4, 6, 8, and 10 are called reference characters or reference numbers. The reason for using even numbers is simply to leave room for other numbers in case you should want to add more description to the drawing later on (and you usually will).

Getting Your Patent

A plastic or rubber lining 12 is placed in the pit for the purpose of holding water similar to the way it is held by pool and reservoir liners. The liner 12 is flattened at the top below ground level 6. A grid of plastic tubes is laid over the liner 12; these are spaced horizontally from each other, uniformly, by 1 or 2 feet; in some cases they are placed from ½ to 3 feet apart. The inside diameter of the tubes is preferably from ¼ to ½ inch but in some cases may advantageously range from ⅛ to 1½ inches.

Note here that I have given a range of dimensions because often dimensional limitations are enough to get a claim allowed by the examiner, which also means, of course, that you will have them as a limitation in your patent when it issues. However, you should be sure to include as wide a dimensional range as you think could possibly work and still be inventively different from the norm. Several times I have gotten claims by putting in dimensional limitations, only to have a competitor copy my product in every detail except for the dimensions, which would just exceed the largest ones I had given. I had no way of knowing, of course, whether I could have received an allowed claim with a larger dimension. Would a judge have ruled it as being equivalent?

The tubes 14 are arranged in single loops, forming U-shaped returns at the far end 16. The tubes 14 are connected alternately to inlet header 18 and outlet header 20 so as to have parallel flow in all of the single loops. Earth is backfilled on top of the liner 12 and lowest tube layer 14, taking care not to damage either one.

A second layer of tubes 22 is installed in the same way as first layer 14 but spaced vertically by the backfilled soil in approximately the same dimensional range as the horizontal spacing. The tubes are likewise connected alternately to headers 18 and 20 to give parallel loops as in the loop grid 14. Soil layer 26 is then backfilled on top of tube layer 22.

This procedure is continued alternately until soil (3) comes within about 2 feet of ground level 6, with the top tube layer 28 being about 3 feet below ground level 6. About 1 foot below ground level 6, a layer of waterproof insulation 30, such as SM Blue Styrofoam as used under ice rinks, is applied on top of the leveled and rolled soil covering the top tube layer 28. Over this is pulled a waterproof synthetic cover 32, which is attached to the liner 12 at the perimeter to prevent rain or drainage water penetration into the cooling storage area 1.

(I realize now that I have not identified the whole purpose of this drawing and thus will introduce reference character 1 by adding at the end of the first sentence of the first paragraph on Figure 1 "to form a cooling storage area 1.")

The cover 32 is then buried with a layer of soil 34 to reestablish the original ground level 6.

Prior to covering headers 18 and 20, connecting pipelines 36 and 38 are run to high airflow fan-coil unit 40 through pump 42 to provide a closed recirculating system through all of the underground parallel tubing loops 14, 22, 28, etc., headers 18 and 20, connecting pipes 36 and 38, pump 42, and fan-coil unit 40.

(It is better to wait to describe the operation of the fan-coil unit and its function until the operation of the whole system is described.)

Vertical fill pipe 44 is installed for the purposes of filling the cooling storage area 1 with water. Perforations 46 in the pipe are used to distribute the water more easily through the layers of soil. More than one fill pipe 44 may be used. These pipes 44 are also used to determine the water level in the cooling storage area 1, which should be filled to about 1 foot below the insulation 30 in order to provide an expansion volume 29 for water as the ice freezes. Since water expands about 9 percent as it freezes, there should be an extra 9 to 10 percent of relatively dry soil to absorb this expansion as the water is squeezed up between the freezing areas around tubes 14, 22, 28, and the others. This will eliminate heaving of the ground level 6 above during freezing operation. Overfilling can be drained away, in many cases, by overflow pipe 45 connected to drain 47 in a conventional manner.

The closed recirculating system described above is now filled with an antifreeze mixture such as 40 percent ethylene glycol and 60 percent water by volume, which has a freezing point of about −15°F, making sure no air remains in the system. Outdoor air thermostat 48 is connected electrically to pump 42 and fan motor 50 and set to turn them on whenever the outdoor air is significantly below the freezing point of water, 32°F. When this occurs, the outdoor cold air is drawn into the fan-coil unit 40 at inlet 52, passes over finned coil 54, and is expelled by fan 56. The recirculating antifreeze mixture is cooled to as much as a few degrees above the outdoor air temperature and is pumped through the ground coils, where it cools the ground and slowly freezes the water in the ground cooling storage area 1.

While more can be said about the operation of the system over the season, this should be left until after the descriptions using the drawings are completed.

For Figure 2, an alternative method should be shown.

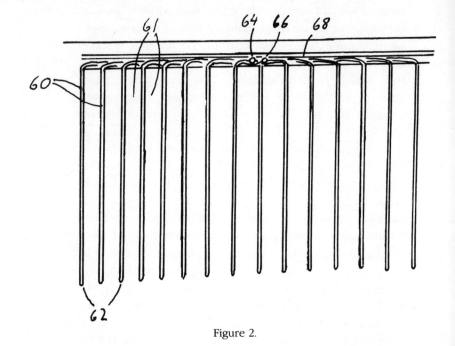

Figure 2.

Alternative ground-freezing system with vertical dual tubes.

You may not always have an alternate method, but, if at all possible, try to put one (or more) in here. Your invention may get protection in one form and not in another. It's hard to tell which method will finally be more salable. Never worry about putting in too much, even if you're unsure of its novelty. My experience is that the contrast between something anticipated in prior art and something with no clear prior reference (your primary method) is helpful. (Note more buzzwords: "novelty," "anticipated," "prior art.")

Figure 2 shows a simpler vertical tube grid in which smaller dual tubes 60 are installed vertically in the ground 61 in drilled, augered, or hammered holes. The lower ends of the dual tubes are connected to each other by the U-shaped bends 62 similar to the ones shown in my Patent No. 4,112,921, Figure No. 2B. Their ends are connected alternately to inlet and outlet headers 64 and 66. These headers are covered by insulation 68 and connected to a pump and fan-coil unit, as shown in Figure 1.

There is no need to repeat what you've clearly explained before.

This embodiment [another convenient word] of the invention is useful where it is not practical to bulldoze a pit 2. No liner 12 is or can be used, so wetting of the ground with hoses (not shown) will be required. Heaving is not likely to occur, since expansion can force water downward or sideward between the tubes into adjacent ground, and the liner is not present to restrain the escaping water flow. It should be understood that the horizontal tube system shown in Figure 1 can be operated without a liner and waterproof cover provided attention is given to keeping the soil in the ice-storage area soaked.

Patent Figure 3 is a schematic diagram of all the possible ways in which the ground cooling storage could work in a system. It shows how the elements relate to each other and when. This helps me and the Patent Office understand it better, adds to patentability, looks intriguing, and shows that I have thought out the systems involved, their practicability, and their analyses. It also introduces a new element: the heat pump.

Figure 3 is a schematic piping diagram, without pumps or valves, of the five modes of operation, labeled A, B, C, D, and E, that occur when a water-to-

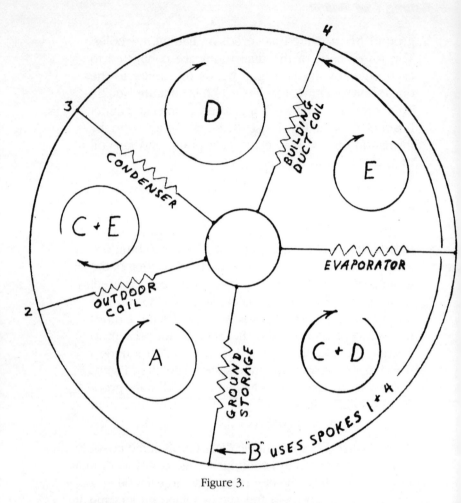

Figure 3.

Schematic piping diagram showing five operating modes.

water heat pump with both liquid-type condenser and evaporator are added to the fan-coil unit and the ground cooling storage system. The five modes are:

A: Winter natural cooling being stored
B: Summer cooling from storage

C: Spring heat-pump cooling being stored
D: Heating from heat pump while cooling being stored
E: Heat-pump cooling in case storage is exhausted

Mode A is the system shown in Figure 1 or Figure 2 in which the natural cold ambient air of winter is being used to freeze ice in the ground. The same antifreeze liquid is recirculated through different paths in all the modes.

Mode B occurs when the "natural" freezing in the ground of Mode A is used to do summer cooling and dehumidifying. A branch parallel circuit from connecting pipes 36 and 38 of Figure 1 must be extended to the duct coil or fan-coil units in the cooling system of the building to be air conditioned. Then space cooling and dehumidification is accomplished by recirculating the antifreeze liquid between the ground storage and the building duct coil, where air in the ducts is cooled and dried by condensation on cold surfaces, as shown by the larger oval arrow labeled B.

Mode C involves the use of a water-to-water heat pump in case there is not enough natural cold weather to provide all the necessary cooling storage. In this case it would be more accurately called an antifreeze-to-antifreeze heat pump, but actually the device is simply known in the trade as a liquid-cooled package chiller as used in chilled-water air conditioning or as sold by Fedders Corporation under the name "Compression Furnace." If the winter season ends wihout the ground being completely frozen in the cooling storage area 1, the heat pump denoted in Figure 3 only by its two parts, condenser and evaporator, can be run on relatively cool nights in the late spring of the year at high efficiency to freeze the remaining ice in the ground. This is not only off the daytime electric peak load but also off the seasonal peak, so that the electric utilities will be benefited and will favor this time of operation. The high efficiency of operation means that at least half of the electricity will be saved for the heat-pump operation compared with that used in air conditioning on hot summer days. In addition, electric rates will be less, and also the part that was naturally frozen will cut costs still further. In southern climates where little or no subfreezing weather is encountered, it will be cost-effective to store all the cooling from the heat pump during seasonal and daily off-peak periods in the coolest weather.

Mode C in Figure 3 is shown by two arrows, the first indicating antifreeze flow recirculating through the evaporator, where it is cooled, to the ground storage where the cooling freezes the ice in the ground; the second arrow shows the heat from the condenser being dissipated to the atmosphere through the outdoor coil.

Mode D is similar to the so-called ACES (Annual Cycle Energy Storage) program developed under Harry Fischer at the Oak Ridge National Laboratory except that ice is frozen in the ground instead of being harvested (or shucked) into a bin or frozen in a tank. Since heating is needed for the buildings, the condenser heat can be supplied while the evaporator is freezing ice in the ground for use in the summer under Mode B. Figure 3 shows these two modes by the circular arrows marked D.

Mode E is a backup mode in case the ice storage is exhausted while cooling is still required. The heat pump would then operate in a conventional manner with condenser heat being rejected as in Mode C and evaporator cooling supplied to the building duct coil or chilled water system.

Getting Your Patent

The drawings have now shown the basic element and system in cross section, an alternative method or structure, and a schematic diagram of the various modes of system operation. Finally, a perspective view of the tubing layout and connections gives a better idea of what it will look like. This drawing should look different from anything seen before and thus add to the novelty.

Figure 4 is a perspective view of the construction of the cooling storage pit area 1 under way. One layer of tubes 14 has already been covered with soil and a

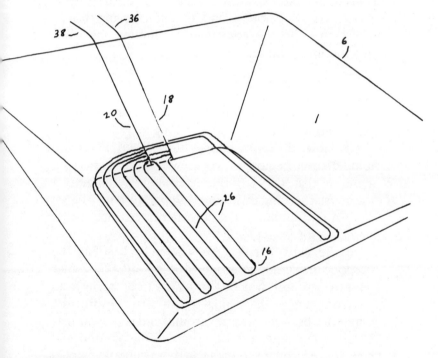

Figure 4.

Perspective sketch of the first layer of tubes in the seasonal-cooling storage pit.

second layer 26 is partially installed, two loops being shown. Headers 18 and 20 are lying on the upper-sloping end of the pit, with some buried tubes from layer 14 already connected, as well as the two new loops from layer 26. Although not shown, there are connector fittings on the headers for all loops from the multiple layers and the multiple loops in each layer. Connecting pipe lines 36 and 38 are shown leading from the headers toward the pump and fan-coil unit and/or heat pump.

All that is left now, before the claims at the end, are various additional explanatory ideas and remarks, descriptions of overall operation, and a windup saying that these remarks shall not be construed as limiting the ways in which the invention may be practiced.

When the ice in the ground is being frozen, the return temperature in the connecting line 36 from the storage to the pump will be very close to the freezing point and will stay that way for a long period of time at a very slowly lowering temperature. When this return temperature begins to go down more rapidly, it signifies that the ground is completely frozen. A thermostat sensing this return temperature may be set to turn off the pump and fan when it reaches such a point, as, for example, 25°F.

In the early part of the winter or late fall, the outdoor air, although above freezing, may be cooler than the ground, which is still somewhat warm from the summer. A differential thermostat sensing the difference between warmer ground and cooler air may be advantageously used to precool the ground and have it prepared for freezing, with no delay when cold weather arrives. Such thermostats, but at different temperatures, are commonly used in solar systems to activate them whenever the solar collector is warmer than the heat storage reservoir.

The building air-conditioning thermostat may be connected to turn on a second pump in a parallel loop going to the building and connecting to lines 36 and 38 as well as turning on the air-conditioning blowers or other parts of the building's cooling system. In this manner, the whole system may become completely automatic, a distinct advantage over those proposed seasonal cooling storage systems mentioned earlier.

In relation to Mode C of Figure 3, there are many areas of the world that would need air conditioning but have only a limited amount of subfreezing weather. In these cases, a relatively small, inexpensive package chiller (heat pump) may be used, operating steadily for the seven non-air-conditioning months to charge up the storage. For example, in a 100-ton design load office building in which cooling is used for 1,000 hours per year, a 10-ton nominal chiller instead of a fan-coil unit operating for 5,000 hours would probably produce the required 100,000 ton-hours; the 10-ton chiller would put out perhaps double its nominal rated value since it would be operating at such low average ambient temperatures. The owner would pay less than 10 percent of the usual demand charge because of the small size and the lower seasonal cost. He would also save on the lower seasonal energy charge, although extra losses in the ground would reduce this saving. The better C.O.P. [coefficient of performance] at the lower ambient temperatures would, however, reduce energy charges.

The cost of the proposed cooling storage area I have calculated to be [always use the personal pronoun "I" for opinions, estimates, calculations, etc.] about $0.30 per ton hour, installed exclusive of the cost of the land, or about $30,000 for the 100-ton building. The 10-ton air-cooled chiller, piping, pumps, valves, and controls would be about $10,000 more, for a total of about $40,000. This is less than the cost of a 100-ton air-conditioning chiller and accessories, installed. Thus the cost is less than or comparable to that of standard air-conditioning equipment, and the cost of operation may be one-half or less.

The "ACES" heat-pump system of Mode D, Figure 3, would require a full-size unit to do the heating; thus it is more expensive than an air-to-air heat pump, which is popular in slightly warmer climates. However, the free cooling may well be a major advantage in certain intermediate, highly populated climates, and the overall cost of operation is considerably less than that of conventional fossil fuel heating with summer electric air conditioning.

The umbrella closing can be just one sentence:

> The above description shall not be construed as limiting the ways in which this invention may be practiced but shall be inclusive of many other variations that do not depart from the broad interest and intent of the invention.

THE CLAIMS

The claims are the legal meat of your protection; they determine infringement and represent the object of all the specifications and drawings. They constitute your invention put into words. This is the toughest part, as far as skill and experience are concerned, and this is where the Patent Office could save everyone much time and expense if it would agree to collaborate in drafting claims. However,

Getting Your Patent 45

there is a shortcut used by many attorneys that makes it easier.

First, decide whether your invention is a structure or a method. A structure is usually one component that can be manufactured and shipped. A method is usually a system, process, or service that is accomplished at the point of use. Often your invention may be both; in that case, it is wise to try for both types of claims if possible.

Second, list all the steps in the method, process, or service or all the contributing parts of the structure. For a method patent, these words may be helpful in starting your claim: "The method of _____, comprising the steps of _____" (follow with your list of vital steps). The first blank space above should be filled in with a phrase beginning with a gerund (that is, a word ending in "ing"), giving a very general name to what is being accomplished. For example, in the case of my invention, it should be "The method of *storing coolness and subsequently releasing it for use at a later time*, comprising the steps of...." The steps should also begin with gerunds, such as "providing," "spacing," "arranging," "forming," "connecting," "adding," "causing," "placing," "covering," and so on.

For a structure claim, this phrase may be helpful:

"A _____ for _____, comprising...." For example, in my case, "A *volume of soil containing multiple tubing loops for storing coolness*, comprising...." is a good way to start. Then follow with your list of nouns spelling out the parts of the structure.

The list should have the bare minimum of steps or parts required for it to work. This would be your broadest claim. Then, if you fear that it might be rejected as unpatentable, you can add as dependent claims other advantageous but nonvital steps or parts.

My list actually may be several lists, because something tells me to claim, first, the storage volume alone; second, the system, including storage volume, outdoor fan-coil unit, and the building cooling system; and, third, the system with storage volume, package chiller, and building cooling system.

My basic list would include the following:

1. An excavated pit
2. A waterproof liner
3. Multiple layers of tubing
4. Multiple layers of soil separating the tubing
5. Supply and return headers
6. Tubing connected in parallel loops to headers
7. Pipe connections to the headers
8. Insulation
9. Covering layer of soil
10. Antifreeze liquid in the tubes

My nonvital but advantageous list would include these items:

11. A waterproof cover sheet to keep excess water out
12. A vertical standpipe for filling with water, measuring its depth, or measuring what percent frozen the storage volume is
13. An overflow drain for controlling water depth
14. Bulldozing with end walls at appropriate slopes

In my system list I would generalize the storage volume so it could also be accommodated by vertical tubes in vertical holes with no pit, no liner, and no insulation, but I would add the outdoor fan coil, pumping means, and building cooling system with necessary piping.

It is hard to know what claims to write but generally eight or more limitations from your list of required parts or steps

Getting Your Patent

is enough. Here we have ten just in the storage volume. We can write claims both on the structure and on the method of creating the structure. Then there are two systems: one for the northern climate with just a fan-coil unit added and one for warmer climates where a heat pump is needed as well.

So here are my six independent claims: three method and three structure, each having storage only; fan-coil system, and heat-pump system. The first is a method claim on constructing the storage only.

CLAIMS

1. The method of constructing the coolness storage volume, comprising the steps of:
   a. Excavating a flat-bottomed pit in the ground
   b. Lining the pit with a waterproof material
   c. Placing alternate layers of U-shaped, looped, synthetic tubing and soil until the pit is nearly full
   d. Providing liquid-supply and -return headers
   e. Connecting one end of each of said U-shaped loops to said supply header and the other to the return header
   f. Attaching the connecting piping to each of the headers for connection to a source of coolness
   g. Insulating the top of the pit below ground level
   h. Covering the insulation with soil
   i. Filling said headers and U-shaped loops of tubing with antifreeze liquid
   j. Filling the void spaces in the soil in said pit with water to a level below said insulation whereby, when coolness is supplied by recirculating a cold antifreeze liquid, the water in the void spaces will be frozen to ice, which coolness may later be retrieved by recirculating a warmer antifreeze liquid

With ten restricting elements of the claim, it is specific enough to appear patentable and yet broad enough to achieve what I need to protect myself.

Here is the storage only as a structure:

2. A volume of soil containing multiple tube loops for storing coolness, comprising:
   a. A flat-bottomed excavated pit
   b. A pit liner of waterproof material
   c. Alternate layers of U-shaped loops of synthetic tubes and soil to near the top of said pit
   d. Liquid-supply and -return headers near an edge of said synthetic tubes
   e. Said headers having connections adapted for connection to supply and return pipes for providing circulating to a coolness-supply system
   f. Waterproof insulation covering the top of the pit below the ground level
   g. A layer of soil covering the insulation
   h. Antifreeze liquid filling said headers and synthetic tubes
   i. Water filling the void spaces in the soil in said pit to a level below said insulation

Third is a method including the storage, the fan coil, and the building cooling:

3. The method of storing winter's natural cooling for use in summer air conditioning comprising the steps of:
   a. Dispensing multiple synthetic tubing loops evenly throughout a volume of wet ground
   b. Providing supply and return headers near the perimeters of said volume of wet ground
   c. Alternately connecting the ends of said tubing loops to said supply and return headers

Getting Your Patent

  d. Connecting said headers to a first set of supply and return pipes and a pump for providing circulating liquid flow to the fan-coil unit
  e. Controlling said fan-coil unit and said pump to blow air at a temperature below the freezing point of water in heat-exchange relationship with the circulating liquid
  f. Forming a coolness storage volume by freezing water in said volume of wet ground surrounding said tubing loops by heat transfer to said circulating liquid
  g. Also connecting said headers to a second set of supply and return pipes
  h. Connecting said second set of pipes to a building cooling system and a pumping means
  i. Circulating a liquid through said tubing loops, headers, second set of pipes, pumping means, and building cooling system, whereby to cool and dehumidify a building when air conditioning is required

Fourth is a system, a form of structure, pertaining to the storage, fan coil, and building cooling:

4. A seasonal coolness storage system for using winter's natural cooling to provide air conditioning, comprising:
    a. Multiple synthetic tubing loops evenly spaced throughout a volume of wet ground
    b. Liquid-supply and -return headers near the perimeter of said volume of wet ground and being connected alternately to the ends of said tubing loops
    c. Said headers having connections adapted for connection to the supply and return pipes for providing circulating liquid flow to a coolness storage system

d. Said coolness supply system consisting of a fan-coil unit, located outdoors above ground, adapted for blowing cold winter air over the coil of said fan-coil unit
   e. Said circulating liquid being an antifreeze solution, which remains liquid in cold ambient conditions and being cooled when passing through said coil
   f. Said wet ground being frozen over time by the passage of the cold circulating liquid to form a coolness storage volume
   g. A building cooling system also connected in parallel to said headers
   h. Pumping means to circulate said antifreeze solution through said tubing loops in said coolness storage volume and then to said building cooling system when air conditioning of the building is required

Fifth is a method claim including storage, heat pump, and building cooling:

5. The method of storing seasonal cooling requirements for a building from a small heat pump operating in other seasons of the year in a building, thereby comprising the steps of:
   a. Same as 3a
   b. Same as 3b
   c. Same as 3c
   d. Same as 3d except change fan-coil unit to heat pump
   e. Controlling said heat pump and said pump to operate and cool said circulating liquid flow during the cooler hours of the year
   f. Same as 3f
   g. Same as 3g

Getting Your Patent

    h. Same as 3h
    i. Same as 3i

There has been a shortcut in the writing of this claim, which will, of course, be typed out in full before filing with the Patent Office.

And last, in a structural system claim including storage, heat pump, and building cooling, a shortcut is again used:

6. A seasonal coolness storage system for using a small heat pump operating through the cooler hours of the year to provide air conditioning, comprising:
   a. Same as 4a
   b. Same as 4b
   c. Same as 4c
   d. Said coolness supply system consisting of a small heat pump located to reject heat to outdoor air
   e. Same as 4e (change last word, "coil," to heat pump)
   f. Same as 4f
   g. Same as 3g
   h. Same as 3h

Writing your own patent is long, hard work, but so is licensing it or starting and running your own business. The patent application comes first, and it can be a real test as to whether you'll be a success as an inventor. If you can somehow afford a good patent attorney to do the whole job, that may well produce a better patent, but it will not make final success any more likely. Using an attorney as editor makes sense in preparing the application, as does using his office as clerk in following it through the protocol. It will be two years or more (in one case it took me thirteen years) before the patent is finally issued. The backlog is large, there is difficulty in hiring enough examiners, the computerizing is incomplete, the cases get more complex each year, and the unnecessary dependent claims get more and more numerous, taking up review time.

FILING

If you file your patent yourself, it is simply mailed to the Commissioner of Patents, Washington, D.C. 20231 (actually in Arlington, Virginia), with the appropriate signed form and the indicated fee. Complete information on all requirements is available in a 60¢ booklet entitled *General Information Concerning Patents*, from the Superintendent of Documents, U.S. Government Printing Office, Washington, D.C. 20402.

Foreign filing is best done by foreign attorneys, but think very hard before spending this money. Filing and translation costs are high and maintenance fees have increased recently to many hundreds or even thousands of dollars per year. It is wise to have an interested licensee or exporter before filing in a foreign country, although the time in which you can do this is very limited. In most countries the limit is within one year of U.S. filing or publication of descriptive material. This means you must decide to file abroad, if you're going to, before you have any indication that you will get a U.S. patent.

One way to tackle foreign patents is to file a Patent Cooperation Treaty (PCT) application, Form PCT/RO/101, with the U.S. Patent and Trademark Office when you file your regular patent application. At that time you select the foreign countries in which you wish to file; most of the ones you would be interested in belong to the thirty-five-member PCT. There is a transmittal fee of $35, a search fee of $3, and a charge of $45 per country. You will then have twenty months (eight more than you would have normally) to select a foreign patent agent for each country and pay that country's fees and translation costs. Also, you will receive a search report from the Patent Office, acting as an international searching authority, that will help you decide finally which countries you want to seek patent protection in. There is also an eleven-member subgroup of the PCT called the European Patent Convention (EPC), which counts as one country for the $45 fee if you file an EPC application as well—but remember that rejection by any one of the eleven results in rejection by all. Again, you should work through an attorney active in foreign patents, since you will eventually need foreign agents, which an attorney can provide.

U.S. patent prosecution chiefly involves responding within a stated time to "office actions," in which the examiner either allows your claims or denies them with an explanation usually referring to patents (prior art) he has discovered that are deemed relevant. If your claims are denied, you may argue for them as filed, amend them, or file new ones, preferably pointing out how you have responded to the examiner's action or opinion. There may also be any number of corrections that are requested in the disclosure or drawings. Order the reference patents, study them carefully, and point out how and why your invention "distinguishes" (a buzzword) from the prior art in ways that someone skilled in the art would not have thought of.

When the examiner agrees that you are entitled to certain claims, they are said to be "allowed." When all your claims are either allowed or rejected, the case is closed and a period is started ending in issuance and printing of your patent, subject, of course, to payment of the final fee. Some people delay issuance as long as possible because protection is for seventeen years from the date of issuance. There may not be any infringers to protect against now, while seventeen years from now a few extra months' protection may be worth a great deal. In any case, the Patent Office tries to hurry issuance along.

Using the phrase "patent pending" on your literature or product nameplate does not give you any legal advantages, but it may scare off a competitor because of the anticipated patent issuance. Generally, though, it is mere puffery and does not have much effect. The important question is always how broad your patent protection will be.

INTERFERENCE

Occasionally, two inventors have similar inventions being reviewed simultaneously and an "interference" is declared. A special board then tries to find out who is the original, first inventor. This is where your diary, with dates and a witness of the entries, is so important. The filing date of the patent application carries a lot of weight and will probably be the deciding factor unless some clear records can be produced demonstrating that the person who filed later actually conceived the idea earlier. Often in such cases, part of the invention may be "anticipated" by the other inventor and part not, so patents are issued on the appropriate parts. If the decision is very close, the earlier applicant may be entitled to copy the claims of the other. If the interference decision is not accepted by the parties, the action may be referred to the U.S. Court of Customs and Patent Appeals, one of the best of the federal tribunals.

INFRINGEMENT

If someone makes or sells or, in some cases, uses a product that comes within the claims of your patent and you can win a suit for patent infringement against them, they may be stopped and/or forced to pay damages or a royalty. The usual defensive claim by the infringer is that your patent is invalid and should never have been granted by the Patent Office. This can be very expensive and take a long time to adjudicate. Recently the Patent Office instituted a new service wherein a challenge to the validity of a patent can be settled by a hearing within the Patent Office for at much less cost than in the usual way. Many attorneys are paying little attention to this service, feeling that in most cases the question will finally have to be settled in court anyway. You will never need a good lawyer more than when you have to explain the technical and inventive features of a patent in dispute to a judge and jury.

CHAPTER 3

FINANCING—THE PART YOU CAN DO

The private inventor may or may not want to be an entrepreneur, to start and manage a business and employ other people. He may not particularly want to run a company, any more than an author wants to be a publisher. But usually he is pushed that way because there seems to be no other alternative. The only way he can get the money to continue product development or begin limited marketing to prove the product's salability is to form a company, assign his patent to that company, and sell shares.

Licensing your patent is theoretically an ideal solution. An experienced company with engineering, manufacturing, and sales facilities would pay you advance money, a minimum payment every quarter or year, and a percentage of each licensed item they sell. You could use the money toward your next invention.

Unfortunately, licensing rarely operates that way. A licensee wants to know that the product has really worked for a certain period of time, that it can be manufactured at the estimated price, that it will sell, and that no unforeseen problems will arise in any phase of its commercialization.

Failing this, he is not going to pay much for it if he bites at all. Licensing in modern times (see Chapter 7) demands more than just a rough working model and a patent application. It usually requires, if you are to receive any substantial return, that you market the product in at least a limited way. This takes money.

THE STING

There are a number of organizations preying on inventors who would try to convince you that it doesn't take a lot of money. Sign up with them, they say, for a fee, and they'll develop your product, prepare drawings, file a patent application, and make every effort to find you a licensee. You feel that your dreams will be answered. Such a company was the Raymond Lee Organization (RLO). Complaints from customers alleged that there were long delays, that RLO's marketing efforts were ineffective, and that status reports were never issued. One of the Better Business Bureaus issued a File Report about these complaints recommending that a customer consult an attorney before signing a contract with RLO involving a substantial amount of money.

RLO sued the Bureau for libel, but the case was later dismissed. In the litigation, it came out that in RLO's work for over 30,000 customers, only three had even made back in royalties the money they had paid out. RLO had carefully honored the terms of the contract, but virtually everyone had felt they were misled. It is estimated that there are some 250 idea brokers doing about $100 million a year in business and servicing some 100,000 hopeful inventors. Although they are not all frauds, many of them are, so be very careful.

Paul S. Shemin, a member of the Special Committee on

Consumer Affairs of the Association of the Bar of the City of New York, discussed this problem at length in an article that appeared in the *New York Law Journal* in 1978. Because the activity is so widespread and so many inventors get hurt, an excerpt from his article follows:

> Invention promoters invest large sums in advertising, usually in handyman's magazines and local newspapers. They offer "valuable" free information to inventors. When an inventor nibbles, the developers send the potential client a package of materials, including several sleek brochures containing seemingly legitimate endorsements by public officials and satisfied clients. Pictures of products ostensibly on the market give the impression of a firm thriving on the successes of its clients. Also included is a confidential disclosure document or record of invention form to be filled out and mailed back to the company. The developer promises to "examine" the invention, a commitment which leads the inventor to expect an evaluation, but which means little more than that a salesperson will soon be in touch.
>
> The inventor is contacted by a company representative who invites the potential client down to the firm's impressive offices, where pictures and certificates adorning the walls create an aura of respectability. A preliminary agreement is proposed in which a market analysis and a patent search will be done, at a cost to the client of $150–$250. While the salesperson is encouraging, the inventor is given the distinct impression that the developer will not go any further unless the preliminary results are positive.
>
> After a month or so, the client will get a phone call (or simulated telegram) suggesting that the inventor hurry down to the promoter's offices. The salesperson sounds excited and notes that the results of the

preliminary research agreement are back. Explicit evaluations are rarely given, even orally, but clients are regularly told that their ideas have great potential, or that the firm has decided to take a chance. One company representative sent a client a personal note—"I would like to discuss these search results and the marketing cost *we* are prepared to assume as a participant" (emphasis theirs). The impression is conveyed that the developer does not make a profit unless the invention is sold or licensed. This is further advanced by the developer's insistence on taking a 10 or 20 percent interest in the invention.

The patent results are presented to the inventor, who of course cannot understand them. The client assumes, incorrectly, that the firm would not go ahead with the venture if the results were not promising. The marketing analysis, little more than a compilation of statistics, contains impressive numbers which delude the unsophisticated client into believing that a detailed dollar analysis has been undertaken.

An example was presented to the Maryland legislature by a Federal Trade Commission official:

> Your product is designed to appeal to teenagers. There are 50 million teenagers in the United States. There are 50,000 teen specialty stores in the U.S. In the first year it is possible to obtain entry to 10 percent of these stores, or 5,000. Each specialty store should sell 100 units making a first year sale of 500,000. The retail cost for your product should be $4. The wholesale cost should be $2. Normally the inventor receives 5 percent of the wholesale price.

The unit and pricing figures are purely arbitrary, but the presentation encourages the inventor to do some simple calculation and arrive at a first year royalty of $50,000.

The total effect of the presentation is designed to convince the inventor that the developer has evaluated the product technically and commercially, that the firm believes the idea to be worth its time and effort, and that if the company's judgment were poor it would not be as obviously affluent as it appears to be. In reality, the product was not evaluated, the firm will expend little time and effort developing and marketing the product, and the company's affluence comes not from selling inventions but from selling contracts.

For $1,000 or so and a minority interest in the invention, what does the shady invention developer agree to do for its client? The firm usually obligates itself to develop the invention, have a patent application prepared, put together a marketing prospectus and undertake a comprehensive marketing and public relations campaign. It all sounds fine, but there's a catch: the development consists of a simple drawing, the patent application is shoddily prepared for a nominal fee ($75–$100) on a mass production basis, the prospectus is a simple one- or two-page nontechnical description of the idea and the marketing and public relations campaign consists of mass mailings to companies and publications with whom the developer has no special contacts.

Because the invention developer has employed this technique for thousands of clients, with virtually no success, the company knows that what it is offering is worthless to the inventor. The concealment of this all-important fact, particularly from a prospective joint venturer, constitutes an omission tantamount to fraud. Although this legal conclusion may be expressed simply in generalizations, proof of fraud of this nature is difficult and requires extensive investigation and testimony.*

FINDING INVESTORS

Let's assume that starting a company seems to be the best option this time. Financing of ventures is divided up into six stages: seed capital, start-up financing, first-stage financing (for refinement of manufacturing and sales), second-stage financing (for growth, even though the product is not yet profitable), third-stage financing (major expansion of a profitable company), and fourth-stage financing (financing as a bridge to going public). Seed capital should come locally from those who know you personally and who believe in you and should supplement money you have yourself. Seed capital is faith money. You cannot expect it to be given as a business investment because, statistically, the odds are too poor. According to William R. Chandler, President of Bay Venture Management, Inc., in San Francisco:

> Typically entrepreneurs tap informal sources (such as family, friends, or groups of private investors), to combine with whatever personal capital they have. I would guess that at least 95% of the successful seed financing must still come from these traditional sources. These informal sources base their investment decisions on their knowledge of the individual and their personal assessment of the business venture.†

---

*Paul S. Shemin, "Consumer Law—Idea Promoter Control: The Time Has Come," *New York Law Journal*, March 30, 1978.

†William R. Chandler, "Pre-Startup Seed Capital," in *Guide to Venture Capital Sources*. Wellesley Hills, Mass.: Capital Publishing Corp., 1981. This article is also available in *How to Raise Venture Capital*, edited by Stanley E. Pratt (New York: Charles Scribner's Sons, 1982).

Venture capital companies are of little use to an inventor at this stage. They don't finance inventors; they finance entrepreneurs. They do finance start-ups sometimes, but, as a rule, only for people branching off from another company who have a proven track record. If you are serious about being an entrepreneur, wanting to build a company at all costs rather than an invention, and have had corporate experience, then you may want to talk to venture capitalists. (See Chapter 4.) Most of us inventors will have to build both our product and our track record as we start up our new operation.

Whether or not you qualify as an entrepreneur in the venture capitalist's sense, you may be about to become one and will find it useful to be aware of the qualities needed for success. In an article in *Guide to Venture Capital Sources*, Alexander L. M. Dingee, Jr., Leonard E. Smollen, and Brian Haslett describe twelve necessary traits on which to rate yourself or have others rate you.* They boil down to drive, self-confidence, persistence, money sense, problem solving, realism, risk sense, failure analysis, use of criticism, assumption of responsibility, use of resources, and setting of standards.

This is a lot to ask of a creative technical innovator who has what the Patent Office calls "a flash of genius." How would Edison have rated? Actually, pretty well. Edison was certainly very strong on drive, self-confidence, persistence, problem solving, failure analysis, assumption of responsibility, use of resources, and setting of standards. But he was nearly broke during his most productive period and much of his work was unrealistic. He paid little heed to criticism. In addition, his heavy investment in iron mining in western

---

*Alexander L. M. Dingee, Jr., Leonard E. Smollen, and Brian Haslett, "Characteristics of a Successful Entrepreneur," *Guide to Venture Capital Sources* (Wellesley Hills, Mass.: Capital Publishing Corp., 1981). Also available in *How to Raise Venture Capital*, edited by Stanley E. Pratt (New York: Charles Scribner's Sons, 1982).

New Jersey was a great, unnecessary risk that went bad and almost sank him. All this is well recounted in biographies such as Matthew Josephson's *Edison* or the sensationalized *A Streak of Luck* by Robert Conot. Edison was a master salesman, as every great inventor must be, but he had much more. The search for the light-bulb filament, for example, reflects the ultimate combination of drive, persistence, and use of present resources; such an "Edisonian search" means that the searcher will try everything, twenty-four hours a day. The lesson Edison teaches us is that a great strength in certain areas overcomes weaknesses in others, which can be compensated for by associates if you'll let them.

The first funds for your invention might come from several different groups of people. Perhaps your friends or relatives will invest more to make it go beyond the model stage. Perhaps you can form a partnership with someone who can give you the strengths you need for the project. Or your best bet might be a wealthy individual who is willing to put up part of his or her money at high risk, placing faith and trust in you as a person. Your investor can well be a retired person or a widow with comfortable means who would not be hurt by a loss. Don't approach this type of prospect unless you are totally committed to your idea and willing to be completely straight about it.

Look for informal investors in the following ways:

1. Advertise in the "Business Opportunities" section of the newspaper. Before April 15, 1982, you were not allowed by the Securities and Exchange Commission to approach large numbers of potential investors without getting SEC approval, which meant obtaining a very expensive registration statement. They would allow you to approach up to twenty people and accept investments from up to ten. However, the SEC has recently issued Regulation D (effective 4/15/82), very good news for

inventors, which allows you to approach and take investments from any number of people as long as the amount you are raising is not over $500,000. Also, brokers can now accept commissions for this kind of work. Other exceptions allow you to (a) take over $500,000 provided all the investors are experienced business people who can judge what they were getting into and (b) take over $500,000 from any group provided not more than thirty-five people are involved and they have been made aware of the risk. Lawyers feel that staying under $500,000 is the safe thing to do because there is no other qualification that you must prove to a court in case of a suit. It is wise, though, to check your local state regulations.

Previously, many people got around the old limit by wording their ad "Partner Wanted," as if they were only looking for one person. Now the way is open for you to go to brokers to find these investors. *Venture* magazine will even run a carefully controlled ad seeking investors for you (see one that I ran, reproduced here).
2. Buy a mailing list of local investors in real estate or tax shelters. These are available from a number of sources; see classified ads.
3. Go to your bank, chamber of commerce, civic organizations, private clubs, money managers, and attorneys. They often hear from individuals looking for investments of varying types.

THE BUSINESS PLAN

To get prospective investors to put up money, you will often need a nicely prepared, typewritten business plan and

For professional investing organizations only, the following investment opportunity is presented for consideration.

COMPANY NAME: **CALMAC Manufacturing Corporation**

DESCRIPTION OF PRODUCTS: Patented thermal storage system for electrical load management manufactured and sold nationally through manufacturers reps, consulting engineers and air conditioning contractors.

AMOUNT AND FORM OF FUNDING SOUGHT: $500,000 private equity investment for 33% of common stock.

REASONS FOR PROJECTED SUCCESS:

Frost and Sullivan says: "The embryonic load management market will explode in the '80s and '90s." They predict $180 million sales for commercial cooling storage by 1989. Booz, Allen & Hamilton in a DOE report predicts $500 million by 1990.

CALMAC's all-plastic low cost modular LEVLOAD Ice Banks make a load managed air conditioning system cheaper than standard air conditioning to buy and much cheaper to operate. Compatibility with standard chillers allows for retrofits.

INDIVIDUALS TO BE RESPONSIBLE FOR IMPLEMENTATION:

Founder, inventor and President of this 35 year old company is Cal MacCracken, holder of 80 U.S. patents and author of Scribner's forthcoming "Handbook for Inventors." John Armstrong, MBA from Harvard, is V.P. and Controller; Mark MacCracken, V.P. Engineering and Manufacturing; Roy Nathan, former National Sales Manager for Climatrol, is Marketing Manager; Wag Waggoner, former Fedders Sales Manager, is Sales Manager; Brian Silvetti is Engineer.

CALMAC has received R & D grants from U.S. DOE totaling over $1 million which have contributed to the thorough evaluation and testing of LEVLOAD Ice Banks.

Further details and a copy of our business plan are available from our attorney to qualified parties. Please contact: Gary Peiffer of Hartman, Lyttle, Coomber & Peiffer, 15 Essex Rd., Paramus, N.J. 07623, (201) 368-1555.

Venture Disclaimer: Neither Venture Magazine, Inc. or any affiliate, agent, officer, director, representative or employee thereof makes any endorsement, guarantee, warranty or representation of any kind whatsoever with respect to the accuracy or completeness of any statement, documentation or information appearing in or relating to this advertisement and the above described company and this advertisement does not in any way constitute an offer to sell or a solicitation of an offer to buy any security of such company. Any interested party must perform and rely on his or its own individual investigation, evaluation and due diligence regarding such company and such person must contact and deal directly with such company in the manner described above in this advertisement.

Sample advertisement for investors, now permitted under SEC Regulation D.

proposal for financing and investment. Not only will your prospects want to study it all themselves, but usually they will refer it to an attorney, banker, or other counselor for advice when you are not present.

Ideally, the business plan might consist of the following:

1. A title page, listing the name of the business, its address, and the names of the principals
2. An abstract, summing up the whole thing in half a page
3. The company: its history, activities, location, and phone number
4. Financial history
5. Capitalization: shares of stock, net worth (assets minus liabilities)
6. Patent
7. Marketing plan
8. Sales to date
9. Sales forecast
10. Channels of distribution
11. Marketing budget
12. Production plan
13. Product design and construction
14. Production process and capacity
15. Current costs and future cost reductions
16. Plant and equipment budget
17. Breakdown of material and labor, pricing and markups
18. Management and their resumes
19. Financial plan with use of proceeds and financial statement
20. Offering (what the buyer gets for his money)
21. Cash-flow budget showing anticipated income and expense for two years

Financing—The Part You Can Do

22. Manpower budget with salaries
23. General and administrative budget (excluding personnel)
24. Product literature and pertinent articles
25. A binder or cover to put it all together

Obviously, the business plan has to be adapted to your particular situation. Putting it together is a good learning experience no matter how many times you may have done it with other products. You get a better perspective of what's ahead and what needs to be done. You no longer live from day to day, seeing just one more hump ahead.

FORMING A CORPORATION

A share in your profits is a vague thing and difficult to sell. Before there are any profits, there are likely to be so many other new investors demanding a priority that the original party is submerged. Forming a corporation by filing with the Secretary of State is a good idea for several reasons: it protects you from personal liabilities; it protects the investor from future investors; and it establishes an entity in the business and legal world that is recognized and understood. The problems inherent in incorporating are the increased paperwork; the costs of complying with all the requirements from state and federal government; and, if your company is very profitable, double taxation both in the corporation and on your compensation. But if you are prepared to deal with these problems you can derive great benefits from incorporating.

Select one lead man or woman from your investor group and have him or her be your salesperson to close with all the others at a meeting or a series of meetings. That person should be on your corporation's board of directors and be the representative of the others. Once you have a "lead" investor, the rest follows more easily.

OTHER SOURCES OF MONEY

Perhaps you have read that the National Bureau of Standards in suburban Washington has an inventor's assistance program. It is possible to get a grant from them after careful screening. I remember reading that only 1 percent or less got help, and yet two men I know well were successful. One had a low-cost solar collector and one a unique heat exchanger. Neither had much hope, both were turned down once, yet both finally got the grants they needed to get started. To find out more about this, write to the National Bureau of Standards, Inventors Assistance Program, Washington, D.C. 20234.

Sources of government money may disappear, but others will spring up. Some states have an office to help sift inventions. They work hard to find a home and help for good ideas in their states. (See Appendix C for a listing of these offices.) Go see the government agencies; at least write. The old adage inventors treasure is "Nothing ventured, nothing gained."

Another possible source of funds is the Small Business Administration, which is headquartered in Washington but has offices around the country. Sometimes they have money and can help inventors, sometimes not. Whey they do, it is usually in the form of a loan guarantee to a bank that would not otherwise make you a loan. See your banker first; unless you have collateral, you will probably be turned down, at which point you can approach the SBA. Keep in mind that the SBA has to make choices too, and may favor certain types of borrowers or companies for political reasons.

Arthur D. Little, a private researcher based in the Boston area, maintains an invention management program that picks about fifteen inventions out of some six hundred per year and tries to get manufacturers interested. However, Little reports little meager success and very few new fortunes.

Robert Onanian started a newsletter for inventors, *Invention Management,* to assist them in finding help, but he couldn't get subscribers until he changed his newsletter's format to appeal to corporations.

What does all this tell us? It simply goes back to Darwin's law of the "survival of the fittest." It is always thus. What percent of aspiring athletes make the pros? What percent of sperm reach the egg? What percent of writers get published? What percent of retail stores survive?

When the excitement and commitment to an idea hits you, who cares about statistics? Who wants to read about what others did? Who wants to use analysis or synthesis? The approach is much more fatalistic. The idea will succeed somehow: it has to; it's meant to.

STARTING UP

The start-up phase usually involves establishing a payroll, running the company, keeping books, paying bills, purchasing materials, using the phone, making drawings, buying tools and instruments and work tables, inventorying parts, writing reports, taking business trips, filling out forms, working with a bank, hiring help, setting up rules, keeping files, servicing equipment, answering mail, and many other things that seem to take you away from what you really want to do. What you must do is take these things one at a time. None is very difficult; it's just the mass of them.

Keep your company simple: go by local state laws with

standard legal forms and certificates, write in broad language as to purpose, use your own name or a form of it as a title, involve as few others as possible, and so on. All your arrangements can be changed later if you succeed in the early stages. If not, why lose any more money than you have to? You'll need an attorney to file your application, but try to get a friend to do it as a favor. There will, however, be a state fee you can't avoid.

You will have to keep a checkbook to list receipts and payments, but avoid starting a payroll until you have to. This will save a lot of paperwork. If people work for you part time, providing a service that they also provide to others, they can be considered consultants or contractors rather than employees. Make sure you get bills from them on their letterheads.

As mentioned before, try to barter or trade as much as you can to keep costs down. Suppliers of materials may sell to you at cost to help you get started. An attorney or engineer may do free work in return for a share of the profits. A college or high school student may provide labor in return for doing a paper or thesis on the experience. A friend may let you have equipment on loan or allow you to use his workshop in exchange for a few favors or a profit share. An employer may let you use the telephone and supplies in return for something he wants from you outside your normal responsibilities.

Later on, in Chapter 8, I'll give specific advice on some of the tasks involved in running a plant. There are many books on starting and running your own business, but priorities are different when an innovative product is the key.

How much money should you try to raise for your start-up? Twice what you think you need to get to a break-even level. Please forgive this rather flip answer, but it is what you hear when you ask the experts that question. They don't know. So many things enter into it. It looks awfully good to have a complete cash-flow estimate for two years on a computer printout, but, as we sometimes forget, the computer is no better than the assumptions you put into it.

It's best to make two estimates: a best case and a worst case—or, better, a good case and a poor case. Put down all your estimated expenses month by month for two years in two columns ("poor" and "good"), and then divide all your income the same way. Usually, the figure you come up with is somewhere in between. That is the figure you double.

How to structure your funds? The best of all is equity and no debt at this stage. In other words, don't borrow money. Let investors take an all or nothing position just like you. When giving out shares in the company, don't give up control at this point, because you will need the equity for future financing. However, no one wants so small a share that they feel they're not really a part of it. A one-third interest is not unreasonable for your early investors.

Be sure you have authorized stock well above that which you issue so there is plenty of room for new financing later on. Don't be too worried about losing control because the smart financiers will let *you*, as the key ingredient of success, have additional shares in one form or another.

The essential person always has control. Whether or not you are the administrator, treasurer, or factory manager is unimportant. The people in those jobs can always be replaced. The person with the technical know-how is usually the essential person provided he keeps the faith.

When I got started, I obtained seed money from a machine shop looking for a product and gave up two-thirds interest. That, I now realize, was too much and unnecessary, but it allowed me to get prototypes operating in the field and to start a small company of six people.

When the machine shop saw that the product I had developed was not suited to them for either manufacture or sales, they did nothing with it; I had to scramble to raise enough money to buy them out. This I luckily did by getting the country's first venture capital firm, American Research and Development Corporation (ARD) of Boston, to invest and buy out my partners through their friends and a few of mine. As soon as ARD said they were coming in, other money was easy to get. In those days venture capitalists were not as wary of inventors as they are now.

So I started with only 20 percent of the stock, but I was in control; thirty-five years later, I still am.

GETTING A PARTNER

Many times in this business the old saw "two heads are better than one" is particularly apt. Here is an actual case of an inventor who took on a partner to handle the sales, and it made him a millionaire.

Fred Ferber was a Swiss machine designer working with the Reynolds Pen Company when they developed the ball-point pen that could "write underwater." He figured that ball-point pens were more suited to low-cost, high-volume usage than to quality applications. So he founded his own company and concentrated on building automated, mass-production machinery. In 1958, when he first took me through his plant, he told me he had the material and labor cost of his pen down to nine-tenths of 1 cent. He sold it to the drugstore chains at 8 cents, and they sold it retail for 25 cents. He had gotten the drugstore marketing and operating capital by taking a fifty-fifty partner, a marketing executive handling over-the-counter items. Much later, Ferber sold his interest for several million dollars and went into nature conservation. His house on a New Jersey mountainside was home to raccoons, deer, possum, dogs, cats, and even a bear. When he died, he was working on a mixture of soil and ground-up garbage to produce a nutrient-growing compound he called Protosoil. He had worked out an automated system whereby cities would be paid for the garbage instead of assuming the costs and problems of landfills.

Ferber's case illustrates that you should do what you do best, which in his case was machine design. Don't try to be all things to all people. Machine design isn't even on the list of entrepreneurial qualities. Ferber certainly had drive, self-confidence, persistence, and problem-solving ability. The partner he took on provided the remaining ingredients necesary for success.

Partnerships will be explored further in Chapter 4.

A PATENT EXCHANGE

When I mentioned the idea of a Patent Exchange in a *Fortune* magazine article (June 28, 1981) on free-lance inventors, I received many letters from inventors wanting to know more about it. Although it is still only in the planning stages, exchange experts are working on it, trying to define the rules to make it viable. The idea is to have a central marketplace where inventors can list their patents and find investors.

Why do we need an exchange, when all an inventor with a patentable invention needs to do is find a licensee, as an author finds a publisher? As explained before, there are far more uncertainties in an invention than in a manuscript, although predicting the success of a book or dramatic work is risky enough. An invention has technical, manufacturing, marketing, patent-validity, and cost unknowns that a literary work does not have. Therefore, until these are cleared up, a licensee is often not willing to pay much for the invention even if it is a good idea.

The Patent Exchange aims to enable an inventor to raise money to develop the product or idea or have it developed for him without (1) putting a value on the invention beforehand, (2) involving the establishment of a corporation, or (3) tying the invention down to one licensee. It would give one or more investors a chance to get in at low cost, ride out the final development and market analysis, and then decide whether to exploit the invention or sell it. It would provide a reliable marketplace where buyer and seller could meet without fear of exploitation from either side. It would give the market a chance to find the product instead of forcing the product to search for its market.

The rules are still being worked out. In theory, however, an inventor would apply to have his patent listed on the Exchange and would agree to certain protections for investors that would keep them from being liable for confidential disclosure or for actions not within the protection of the patent. The inventor would declare, in a prospectus to be distributed, how many shares would be established (which he would initially own) as well as the offering price and complete background information. He would then be free to sell shares as he wished, realizing that ownership of one or more shares gives the buyer a free right to use the patent but not to license it to others. The buyer could develop the invention himself, or the inventor could do it with the money from the sale of the shares.

If the development were successful, the value of the shares would go up. Others could buy shares either from the inventor or from the original or subsequent investors. The inventor could choose to hold on to one or more shares so as to sell at the best time. No licensing or royalties would be involved. The free market and the true value of the idea would provide a just reward. The inventor could get his money and not be involved in the operations of the company.

I'm sure there are other possible formats for the Patent Exchange. The essential thing is to bring the independent creators in this country together with the people who can commercialize their ideas. Nothing could do more to reduce unemployment and raise our general economic well-being than to facilitate these sorts of partnerships and investment opportunities.

CHAPTER 4

FINANCING—
THE BIG LEAGUES

Unless you can get by on $250,000 or less from private sources, chances are you are going to have to learn some sophisticated financing techniques. The odds for success of your invention are far better if you can raise the money you need. Instead of 1 in 10, they rise to 1 in 2 or 3 when big money is involved. The success rate is much higher than it used to be before this sophistication began and, no doubt, it will be even better in the future as better methods of forecasting, analysis, and management are developed.

The major obstacle to obtaining big money for inventions is the lenders' desire to follow trends. Because of trends, many poor investments are funded and many good ones ignored. Trends are exaggerated by laziness, the tendency to ignore the "good old-fashioned hard work" John Houseman talks about in the commercials. Trends cause mistakes in judgment, oversaturation of particular markets, and recessionary retrenchments as burned feathers heal. As Frank Capra, the famous movie director, said when honored on television, "My advice to young movie makers is *never follow trends; start your own.*"

Financing—The Big Leagues

The big leaguers in venture financing (any new business is called a venture) are the venture capitalists, the large corporations, the R&D partnership tax shelters, and the underwriters who sell stock to the public. Remember that these people know money like you know your specialty. Since they live with it, they know what it can do and how easily it can be lost. Just as you wouldn't think of omitting a key bolt that holds your device together, they have certain bolts that hold their money machines together, and they won't tamper with them.

VENTURE CAPITAL

Venture capital firms, which sprang up as regulated Small Business Investment Companies (SBICs) in the early sixties, were not as active in the late sixties and early seventies, but now they are proliferating more than ever. They are more knowledgeable, they have had far more experience with what works and what doesn't work, and they know how to assess a specific project. They know what fields they want to be in, how large an investment they want to make, and what stage in a company's development interests them. Unfortunately, they are less willing to take chances than they used to be, and they often syndicate deals among themselves to further lessen their risks.

The *Guide to Venture Capital Sources*, previously mentioned, has extensive listings of venture capital companies throughout the United States and Canada, arranged by state or province with fields of interest (industry reference), size of investment, and type of financing tabulated. Geographical limitations are also given, but often these are more flexible than other requirements. The same book has excellent articles on all phases of venture capital. It also lists stock underwriters for small-company public underwritings.

Inventors often find it difficult to get backing from venture capitalists, who see numerous risks in their kind of enterprise. As Professor William Wetzel of the University of New Hampshire's Whittemore School of Business and Economics puts it, "The odds of picking a winner are slim, down-side risks are close to 100 percent, relatively small amounts of money are involved, the costs of investment supervision and guidance are high, and the length of time between investment and potential cash recapture generally exceeds the exit horizon of venture capital firms."

However, there are a number of things you can do to make your proposal more attractive to venture financiers. First of all, don't try to sell them over the telephone. Venture capitalists are not really interested in ideas, patents, and inventions; they prefer to work with businesses. Submit in writing a complete business plan, financial statements, and proposed budgets. Also submit the names of your suppliers and customers, so they can find out what they need to know about your product and its market (they will ask for these names when they start investigating your business anyway, so you might as well make things easier by supplying them right up front).

If you don't need at least $250,000, don't bother approaching a venture capital firm. It doesn't pay for them to lend less than that because of their high overhead. They will want to get their money back, either in cash or stock value, in three to four years—a relatively short period even for most successful inventions. They will probably want an equity position of more than 20 percent and less than 50 percent. Fortunately, most venture capitalists don't ask for control, because that would remove the entrepreneur's incentive. They will, however, insist on strict budgets, monthly meetings with the principals, and restrictions on executive salaries, expense accounts, and capital expenditures.

Finally, venture capitalists prefer to work with partnerships, particularly partners who have been in business together before and have demonstrated compatibility and stability.

The fact is that the outlook for inventors trying to start a new business all by themselves, obtaining financing other than from friends and relatives, is dismal. Venture capitalists don't like to become involved with a company unless it is managed by a team of people who are strong in all the important areas. And they want to know that there is backup in case something happens to one of the principals. One of the most important things an inventor can do, therefore, is pick up a partner to help commercialize his idea and make his company a solid, professional business that will be of more than passing interest to venture capitalists and other big-league financiers.

THE PARTNER

There are literally millions of men and women with experience in small business. Many of them would be glad to have something new and exciting in their lives, something on which they could possibly build an equity without investing anything except time, and without risking their bread-and-butter business. Such a partner can bring you the things you most need: credibility, business management, experience, local recognition, judgment, established procedures, office help, reduced expenses, and a proven track record.

Probably this partner should not be the development company discussed in Chapter 6. Business promotion and management are quite different from product development. Also, developers are concerned primarily with the problems in your idea, while the business partner should focus on the strengths and potential of it. Of course, the business partner must be as mindful of the technical limitations as the developer, who is also in business to make money, is aware of future possibilities. But each should do what they do best.

Let's take a fictitious example. Your idea is a table centerpiece, made of glass, that holds flowers, a floating candle, water, a sparkling powder in the water, a pump operated by a photovoltaic cell from the candlelight, and a small fountain. You've made a rough model and it works. Friends whose opinion you trust have told you it will sell very well as a $50 gift. You have a good eye for what makes an attractive gift, and you have ideas for other gift items. You filed a patent application on your idea based primarily on the fact that the flame of the floating candle stays in about the same place as it burns down and thus can continue to power the photovoltaic cell and pump over a long period.

You went to a bank for financing; they said no but sent you to a venture capital company. The venture capitalists were painfully polite but said they were looking for something with a much bigger potential than a single gift item and why didn't you license it to a gift manufacturer. You called one and he said that he really didn't see any market for it; besides, it would cost much more than $50 retail, because markups in the gift business were such that it would have to cost only $12.50 to manufacture in order to sell at $50.

You had the invention operating as a centerpiece on your dining room table, all decorated with fresh flowers, when your aunt came to dinner. She particularly liked the way the flowers sparkled as they were wet by the fountain spray and the sparkling powder. All of a sudden you thought it might be a good way to sell more flowers, and the florist in town seemed like the man to approach.

When you talked to the florist, it turned out that he was active in his trade association and civic groups and had an interest in a delivery company that made the rounds for many of the local merchants. He liked your device and got quite enthusiastic about the many ways to use it with flowers and even hydroponic plants.

After thinking it over, he said he'd be interested in being your partner on a fifty-fifty basis, keeping it separate from his business but assisting you in every way that he and his business could. That seemed steep to you, for, after all, you had done a lot of work and he had done nothing. Still you had no other alternative and you could put a clause in the partnership agreement stating that if no outside money were raised within one year, all rights would revert to you. If financing were obtained, the partnership would be dissolved and a corporation formed.

And so the partnership was begun. The florist's lawyer drew up the agreement at no charge to you. The partnership called itself Floral Power and Light, hoping to attract attention, sound bigger than it was, and cover other similar types of products. You had thought of plug-in lighted flower stands, plug-in waterfall and plant combinations, and even fish tanks with floral displays.

Your partner, John, brought the invention up with his florist group and you were invited to display and describe it at their next monthly meeting. This went well and, from suggestions given, a line of products was worked out on paper. John was so enthusiastic that he agreed to lend the partnership $10,000 so you could make up the prototype products; he, in turn, wrote up a product literature sheet describing the products with model numbers, drawings, and even prices. He also, with the aid of his banker, made up a budget, a cash requirements and flow sheet, and a profit-and-loss estimate. You wrote up the history, the product and market description, the marketing and sales plan, and the other parts of the necessary written proposal described in Chapter 3.

Armed with all this, you went to a venture capital company in your area that you had picked out of the *Guide to Venture Capital Sources* as being interested in the manufacture of retail products, start-up operations, and $100,000 to $300,000 minimum investments.

There were questions on the strength of the patent, the cost of manufacturing, and the use of funds. Mostly, however, the venture capital company's concern was centered around the market for the product. John's figures on the total sale of flowers, assuming that 1 percent of those buyers bought Floral Power and Light for their table, put the market well into seven figures. The venture capital company was interested.

They requested a few more models for "market research." Actually, that meant taking it home to see what their families and friends thought. You were smart enough to deliver these to the homes in person, set them up, decorate them with the aid of John's best flower arranger, and then demonstrate exactly how to operate them. This, of course, was not very good market research, but it was good public relations—it worked; it was time to consider personnel management, a manufacturing operation, and, finally, the financial deal.

In all this, John was invaluable. His references gave good reports of character and community standing. His demonstrated mangement skills, selection of personnel, and ability to work with people was persuasive. However, his lack of manufacturing experience and only part-time availability became a major stumbling block.

John sought out a local manufacturer to become a board member of the new corporation in return for 5 percent of stock interest. In addition, he agreed that the new compa-

Financing—The Big Leagues

ny's operations could be set up on some vacant land adjoining his greenhouses, so he could give the company a lot more of his time and let his son run more of the florist business.

The deal finally went through at a total of $250,000 invested for 35 percent equity and a long-term low-interest subordinated note for $200,000. The note, the investors said, being subordinated to other lenders, would not affect bank borrowing and would allow them to be repaid without paying taxes. The corporation was formed. A five-member board was appointed; it comprised you, John, the manufacturer, and two of the venture capitalists. You, as president, were in business to sink or swim.

The strikes against you at this point are several:

1. $250,000 will not get you very far from a start-up. More money will have to go in within two years, which will dilute you further from your present 30 percent.
2. The long-term debt (note) will become a real dead weight when future money is needed.
3. The hiring of good sales and manufacturing people is difficult in a start-up.
4. You have to find or build a plant, develop the product, plan the marketing and all its features, and start selling within six months to meet your budget.

As it turns out in our example, you did need more money in two years and the venture capital firm, knowing it had no other choice than to write it off, supplied $100,000. But it took 60 percent ownership of your company, leaving you, John, and the manufacturer with the rest. Fortunately, the notes were converted into preferred stock so you would look better in the eyes of the bank when you need a loan.

Close rein was kept on you by your board of directors, your small salary was made even smaller, and the few employees you had became fewer. Then you wangled your first big order, and things started looking up.

We'll describe later how manufacturing and sales can be handled.

CORPORATE FINANCE

Many corporations have venture capital departments that finance small companies with some relevance to the parent company. They operate in virtually the same ways as the independent venture capitalists except that they have the overriding interest of the corporate image to protect.

For example, most major oil companies have invested in solar companies so as to build their image and keep a finger in the pie in case solar does well. They are willing to lose a little money at it, but when Exxon Enterprises found they had lost $30 million in Daystar Corporation, making copper-based solar collectors, they sold out fast at a huge loss. Having been burned and embarrassed, they refuse now to consider any alternative energy or conservation proposals.

For the inventor who wants to ensure his independence, the danger of acquisition by a corporation that has invested in you is very real. There are many ways a large minority shareholder can control the future of a small company that is almost certain to need additional financing at some point.

The good aspect of this type of arrangement is that better understanding of your product and its technology and marketing may result from the similar interest that brought you to your investors in the first place. They may not be as quick to reject it in response to a trend and may be more able to give you prompt answers.

THE R&D PARTNERSHIP TAX SHELTER

Tax shelters (investments that are deductible from personal income tax) have grown in the last few years, largely in the fields of real estate and oil and gas drilling. More recently the R&D (research and development) partnership tax shelter has entered the finance scene for new products. The key thing common to these tax shelters is that the investment is "at risk," with none of the guarantees or built-in protections that venture capitalists strive to get for their after-tax money. To utilize before-tax money, the investment must pose a true risk: if the R&D or its commercialization is unsuccessful, the investment—except for the initial deduction—is lost.

The unique aspect attractive to high-income investors is that if all goes well, the R&D work will result in a patent that can be licensed; the incoming royalties can then be treated as long-term capital gains. The present law taxes capital gains at 40 percent of the regular income rate. To qualify for this, there must be more than one year between "reduction to practice" of the invention (making an operating prototype, in patent-law terminology) and receipt of income.

These tax shelters are usually structured and initiated by CPA firms and tax lawyers who have clients in the 50 percent tax bracket or by public offering specialists who shift to the most popular financing method. What they look for is a marketer or inventor with a good track record; this will help them sell their investors, who usually have little technical appreciation of the R&D process. These financial organizers become known as the general partner.

So if you have established a good reputation on your own, contact your local CPA or tax lawyer and find out who handles tax shelters. Ask them to study up on R&D partnerships and read the articles on this subject listed in Appendix H ("Recommended Reading") at the back of this book.

If your reputation as an inventor is yet to be made, ally yourself with a well-known marketer. You would be the R&D performer (or you could subcontract or supervise the work) and the marketer would be the licensee, paying royalties to the investors. You, as inventor, would have a percentage plus payment for your part of the R&D. If there is no predesignated marketer, a licensee will be sought when the time comes.

Perhaps independent groups—such as technical professional or engineering societies, government departments or laboratories, institutes, test laboratories, or universities—will begin evaluating inventions. Then, with a positive evaluation of your invention by a reputable group, it may become easier for you to work directly with your general partner.

Although a formal public offering registration with the Securities and Exchange Commission (SEC) is not required for the R&D partnership, the paperwork is still very involved, and a prospectus with all the usual SEC warnings and cautions must be filed with the state attorney general or other cognizant authority. Costs of filing are high, so projects should have a minimum practical investment value of perhaps $250,000.

It appears that this new way of putting investor and inventor together will grow rapidly. Everyone benefits. It is vital, however, that this be handled intelligently and not become another instance of "hype," where the fee is more important than the product.

GOING PUBLIC

Another financing alternative is a public stock issue, through which you might raise a good deal of public money but would also become a public corporation with all the responsibilities, the SEC reports, and shareholder relations that this involves.

To make a public offering, you will need a fairly prominent person as your partner (someone more prominent than John the florist in our previous example). It's best to work with someone who has been involved in substantial corporate management or finance when attempting a start-up underwriting.

Usually a company must be worth at least $1 million before an underwriter will consider it, and $5 million is average. In 1980, the year when ventures boomed, 135 firms worth less than $5 million had public offerings and 50 worth between $5 and $10 million. If you have a product or service that really hits the public imagination, if your company has a strong leader, if you have the money to pay the fees, and if your timing is right (as it would have been in 1980), a public offering may make sense.

A public issue takes about six to nine months. Venture capital can usually be raised in three to six months. Individuals with private capital can be solicited in one month. You may find that the amount of money you receive is roughly proportional to the time and money spent. The main lesson is that it takes hard work and persistence.

Peter Wallace has outlined the pros and cons of going public.* On the plus side he lists the following:

1. You give away less stock for more money.
2. It's easier to get more money later if all is going well.
3. Investors and founders can get cash.

---

* Peter W. Wallace, "Public Financing for Smaller Companies," in *Guide to Venture Capital Sources*. Wellesley Hills, Mass.: Capital Publishing Corp., 1981. Also available in *How to Raise Venture Capital*, edited by Stanley E. Pratt (New York: Charles Scribner's Sons, 1982).

4. The objective valuation allows for management remuneration, employee motivation, and means for acquisition.
5. It enhances the company image and increases availability of loans and credits.

On the negative side, he lists these points:

1. It requires an itemized disclosure of all business details and history.
2. It requires expensive quarterly reports and communications.
3. Smooth growth patterns are needed to avoid stock price fluctuations.
4. The cost of going public is $150,000 to $200,000, even for a small company.

There is a general feeling that it is not wise for an inventor to go public until the company has gone beyond the invention stage into the business growth phase with profits to show. An inventor has a much better chance of surviving in a private company funded by individuals or venture capital firms than he does in a publicly funded one. But there are always exceptions.

The new SEC Regulation D permits a form of going public handled by a "penny broker" without a formal registration statement but with a total investment under $500,000. This form of super-risk investment necessitates big rewards and may strip the inventor of his patent if he doesn't make it on that money. However, it's a new, open ball game for inventors; this should stimulate creativity, more successes and more failures, and more employment with economic recovery.

If you're very interested in big league ventures, you should subscribe to *Venture* magazine (35 W. 45th Street,

Financing—The Big Leagues

New York, NY 10036). The May 1982 issue had articles on the top 100 companies with the founder still active in management and on the top 50 firms founded in the last ten years. Although the trendy computer group was most common among the 50, with the famous Apple Computer ($334 million in sales) number two, number one was a food packer, Keystone Food Corporation of Bryn Mawr, Pennsylvania, with $419 million sales; this older company was reorganized and brought out anew under very careful planning and direction. Two former directors of my company were instrumental in this. Oil and gas drilling, real estate, and health care products also figured prominently in the list.

All these companies had dynamic leaders and were developed through long, hard work and good planning. They had a product or service that fitted a market which they anticipated accurately. It is extremely difficult for an inventor to do this by himself, but for a team it can work. Dr. Edwin Land, chairman of Polaroid, is the best example I know. As chairman, he guided his company's direction and represented it very effectively in dealings with the shareholders and public. But he left the day-to-day operation of the company to others while he headed the research and development that was so vital to its growth.

CHAPTER 5

DEVELOPMENT ON YOUR OWN

Patenting and financing are steps in the inventing process, but the key to it all is the product, process, or service itself and its refinement into a practical entity. Research, design, and development (RD&D) are the three technical phases essential to every successful product. Research is gathering the knowledge; design is assembling the knowledge; and development is arranging the knowledge.

Products are subject to the same laws of "survival of the fittest" as are plants and animals. However, their evolution is compressed into the process called development, covering the period from prototype to final design. It is usually a laboratory process, involving testing, changing, retesting, redesigning, rebuilding, and so on until a suitable product begins to emerge. It is the part of the job that Edison labeled "98 percent perspiration."

When I first worked for General Electric Company as a test engineer in 1941, shipboard instruments for the Navy were being developed in the Industrial Controls Department. These were required to withstand a direct hit on the ship by an enemy shell. Previously, a direct hit would have destroyed the instruments and sent the ship out of control. I operated a shock testing machine that had a heavy vertical steel plate mounted on springs, with a 400-pound steel hammer pivoted to swing up 5 feet and then swing down and hit the plate. My job was to mount the meter (the instrument being tested) on the front side of the plate, swing the hammer 90 degrees 5 feet above the plate, and release the hammer, thus giving the plate a 2,000 foot-pound shock.

Down the hammer would come, hitting the plate with a tremendous bang. The meter would break into a hundred pieces, which flew all over the room. The dejected project engineer would pick up all the pieces carefully, put them in a box, and literally trudge back to the drawing board.

Next week he'd be back with a new arrangement. I'd smash the hammer down again. This time only fifty pieces would fly out, and he would pick them up with no sign of emotion. The following week it was down to ten pieces. On the last day of my three-month assignment as a shock tester, the engineer appeared for the umpteenth time. I mounted his meter, drew up the hammer, and let it fall with its usual deafening noise. And nothing happened. Not only did the meter stay in one piece, but when he switched it on, it worked. The engineer didn't jump for joy, as in the movies. He just smiled and said, "That's better. This should do the job the Navy wants. I'll be bringing down a relay tomorrow." I used to wonder during the war that followed soon after how many lives those shock-resistant meters helped save.

Development is part analysis, part synthesis. General Electric found it took both analysts and synthesists to do a given job, and they had training programs for both. The analysts are the mathematicians, in those days behind a slide rule, today at a computer simulating the problem in silicon chips. They carefully pick things apart. The synthesists work by feel or hunch, by innovation and by trial and error; they put things together. Each is essential to development.

I am disturbed at the present tendency of bright young engineers, up to the age of about thirty, to concentrate only on high technology and computer modeling, leaving hardware development to the older group. There will soon be a gap in talent for the more practical products. Computer modeling and algorithms have their place as tools; they are, however, fantasy, not the stuff of real life.

Doing the development on your invention yourself or with a small staff can be most instructive and valuable. It tells you exactly where the problems in your product lie and gives you the chance to be involved with every aspect of its creation. However, it is also enormously time-consuming and frustrating; many people do not have the temperament or resources for it. If you are one of these people, Chapter 6 explains how to go about finding the right group to handle the development of your invention. For those who want to do it themselves, here are some suggestions to help you along the way.

THE DEVELOPMENT PROCESS

The key to efficient development is in the planning. Assume along with Murphy that "Everything that can go wrong will." Seek out as much help—free help—as possible. There are thousands of suppliers of every type of material and component available to give that help. Your job is simply to find and talk with the right people.

Use the *Thomas Register*, directories from trade magazines, advertising, the Yellow Pages, and catalogs. You can save time and money by sending a form letter to a whole list of suppliers in all parts of the country, asking for their

Development on Your Own

literature and the addresses and phone numbers of local representatives. If the literature looks promising, call the rep. If the company seems to have a product that can help you, go see its engineers. What you're looking for is advice on all phases of how to use its product, what problems to watch for, costs, whether its engineers have seen anything like your invention before, and so on. To get the most information, take the most helpful person to lunch.

Hiring a Staff

If you have raised some money and can hire one or more employees, take time to choose the right people. I think hiring has been my biggest weakness over the years. There is a tendency to be so impatient to get going that you take the first person you like and make excuses for any lapses or weaknesses.

There are a number of ways to learn hiring techniques and good books on the subject are available (see Appendix H, "Recommended Reading"). I want to mention only two important maxims: check each person's references carefully and remember, if you're not happy with the employee, there's always somebody better out there.

To help single out the creative thinkers, we give a little test of puzzle problems that I learned from Theodore Edison long ago. For example: how can you connect nine dots in the form of a square, three rows of three, with four continuous straight lines? (The solution is on page 94.)

Or show how you would wire one electric light to go on and off from each of two switches, such as you would have at the top and bottom of a stairway. The key things to watch for are whether an applicant enjoys trying to solve problems or is dismayed; whether he tries many things or sticks to one approach; whether he can concentrate with distractions around; and whether he understands the instructions quickly and follows them.

DESIGNING FOR PRODUCTION

When you switch from one-at-a-time construction to "volume," "mass," or "quantity" production, a completely new design is needed, one that takes into account the materials and processes of production. Materials include metals, plastics, elastomers, finishes, adhesives, lubricants, organic and inorganic chemicals, semiconductors, insulators, and so on. Processes include turning, milling, braking, rolling, grinding, shearing, drilling, lapping, knurling, stamping, forging, arc welding, spot and seam welding, heliarc welding, extruding, casting, molding, and many others. It would be very wise to learn about these if you aren't already familiar with them, since you cannot design for production if you don't know how things are made. A good book to consult is Roy A. Lindberg's *Processes and Materials* (Boston: Allyn and Bacon, 2nd ed. 1977).

Designing for production starts at the drawing board with a layout—a large sheet of paper on which faint, easily erased lines are tentatively put down to get the feel of what

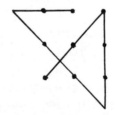

Solution to the puzzle on page 93.

the product will be. It bears only a slight relation to the prototypes you've made before. The scale is usually adapted to fit the space. A drafting machine, which draws lines that are straight and parallel, is a great help; such devices are available in a variety of prices and sizes.

If you haven't had this kind of training, consider taking a drafting course in your spare time. These are given in adult education programs, in vocational courses, and at many design schools. Drafting is fun and fairly easy to learn; it vastly improves your ability to design.

Your layout is in some ways like an artist's canvas. It is something to be played with, viewed from all angles, and changed again and again. A lot of time is spent on small areas of it in the effort to get it right. It differs most from an artist's creation in that many people will often stand around it, discussing and improving and figuring out how each part will be made and assembled.

When the layout is set, a perspective drawing or, if possible, a mock-up is often made to see how the thing will look. Take the drawing or mock-up to suppliers to get their advice on how the product should be made. They are the best source of answers to design problems; for example, they might tell you that aluminum or plastic extrusions, squeezed in endless lengths like toothpaste and in almost any profile, can, if they can be worked into the design, reduce both tooling and per-piece costs. Extrusions add strength, weigh very little, are low in cost, and look modern. As mentioned before, check the *Thomas Register* in your library for the names of these suppliers.

The next step is making detailed dimensional drawings of each part and sometimes of the stages required to make a part. You must always consider the sequence of manufacturing when you work out your dimensions. If you draw a piece of sheet metal in which holes are drilled or punched in the flat and then plan to bend that sheet metal in a press brake, the holes may not line up as you expect. Often the best approach is to bend up a piece without holes, then put the holes in and flatten the piece back out. This will then serve as a template for cutting that piece. Clothing patterns to be cut in the flat before the material is sewn are also designed in this way.

Always include a tolerance in the dimensions used in manufacturing processes; it is difficult to be perfectly accurate and a miss can be costly. Tolerance, which is usually indicated as plus or minus a decimal or fraction of an inch, should be as loose as possible for low cost, but remember that adjoining parts affect each other. If all the parts in one direction are made at the upper limits of the tolerance, the total length may vary by the sum of all those tolerances. So, when selecting tolerances, start with large assembly dimensions and then come down to the parts. Make sure they add up—a procedure called "stacking of tolerances."

When GE developed the first jet engine selected for quantity production in 1944, a year before the end of World War II, the Air Force told us to turn it over to Chevrolet. We called it the TG-180 (for turbine gas), but it was renamed the J-33. Everything else about it was changed too, it seemed. I had thought we had good drawings, but the Chevrolet designers quickly showed us that our tolerances didn't add up, our dimensions were incorrectly referenced, that there could be interferences between parts made within the tolerances we gave, that suppliers could take advantage of us, and so on. We were chagrined, but the head of sheet-metal design in the Chevrolet central office in Detroit, said, "Oh, that's nothing. We had to make thirty-two thousand drawing changes when we took over making the R-2800 radial engine from Curtiss-Wright." Automobile makers are superb at production design. They have to be. And we all know that even they have their lemons.

LABORATORY TESTING

There are two elements of testing: trying and relying. Trying out a number of things in the hope that one will work, at least for a while, is often called jury-rigging; it's very different from reliability testing. I was once involved in a jury-rig that saved a great deal of time, when I worked in Tony Nerad's section of the GE Research Laboratory (called the House of Magic), during World War II. As always, Tony was on a crash program, his assignment this time being the development of a combustion chamber for the then top-secret jet engine. He had found that the answer was a sheet-metal cylinder through which oil was sprayed. One of the cylinder's ends was closed and it was to have a pattern of holes for the air supply—but what pattern? Tony turned that question over to me. After several tries with models that each took several days to build, I decided to partially punch hundreds of holes in the cylinder but leave them closed, like knock-outs in the side of an electrical box. Then I could open or close them as I wished by bending the slug, without disturbing the test setup or involving a shop. We made fifty tries in three days and came up with the basic pattern that has been used ever since.

You must be as ingenious in devising your test apparatus as you were in your inventing. Testing can soak up tremendous amounts of time and money and may or may not give clear answers. There are so many types of tests you could run for each product and so many millions of possible components that you will have to find the answers yourself, but there are places to go to get help.

Professional societies in engineering such as ASHRAE (American Society of Heating, Refrigerating, and Air Conditioning Engineers) and ASME (American Society of Mechanical Engineers) issue testing standards. The ASTM (American Society for Testing and Materials) in Philadelphia also has vast numbers of such standards available. The National Bureau of Standards (NBS) in Washington has publications on almost every subject pertaining to testing and measurements. The NTIS (National Technical Information Service) offers, for a fee, current publications in many fields that are of vast help. (See Appendix H, "Recommended Reading," for the addresses of these groups.)

These standards may apply to testing an entire invention, such as a new solar collector, or they may apply only to components being used, such as the motor and blower in a respirator. Many tell you exactly how your tests should be run, what should be measured, and even what equipment should be used.

SGA (formerly Scientific Glass Apparatus) puts out a catalog for chemical and physics laboratories; Honeywell has a catalog on electrical controls; Radio Shack offers many types of electronic and electrical measuring devices. In the major cities, you'll find stores where used test equipment can be bought or rented. Service companies also have many types of testers and can tell you where to get them.

FIELD TESTING

When people think of testing, they usually think of the laboratory process, with controlled conditions and instruments that measure performance. This is, of course, an essential step; however, the most important and most ignored aspect of testing is that which goes on in the field, where the product is in the hands of the consumer or buyer and is being given everyday use.

As mentioned before, Murphy's law, "Whatever can go wrong will," applies especially to product development. I'm not sure who Murphy was, but I picture the inventor of the Murphy bed, having told his wife that the bed was perfectly safe, stating those now-famous words as he watches the bed fold up into the wall with her on it.

All of us inventors have been in that kind of humiliating position many times. Consumers will uncover more ways for something to go wrong than you could ever imagine. GE's Monitor-Top refrigerator, which ranks with the Model T Ford as one of the most successful products ever made, initially had a serious flaw in its modernized version: the hole for the electric cable was just big enough for a mouse to crawl into and build a warm nest. For some reason this was not a popular feature, and that version of the Monitor-Top sank.

In a good field test, the buyer should pay something for the product, since freebies are not regarded as critically as are purchased items. If you want to call it a market test when actually you are most interested in finding out how well the product works, that's fine. Don't hang around and coach the buyer, though, or you'll lose the value of the test.

Try to get letters from buyers describing what they liked and disliked about the product. Later on, you might engage a marketer or licensee who, while knowing how to cure certain flaws easily, might, at this early stage, worry needlessly about features that buyers actually mention with approval in their letters. These letters will also show that you have made a concerted effort to do your best, which might protect you from a charge of negligence. Be sure that you have a product liability insurance policy to cover you if there is any possibility of an accident, such as fire, explosion, or leak. Even if your product is not at fault, it will be immediately suspect because it is new.

If you find you have a marginal design, one that works 95 percent of the time but has occasional problems in spite of all you can do, it is best to face the situation squarely and start the design over. This is one of the hardest decisions in inventing; it's like abandoning a child.

Don't let an unreliable secondary feature drag down a good primary one. Inventors often try to work too many ideas into one design. The product shouldn't be more complicated than it needs to be—the odds against a good invention succeeding are heavy enough and shouldn't be compounded.

COST ESTIMATING

Part of the development function is preparing a bill of materials, a list that includes the material specification, blank material size, sources of supply, and other descriptive and ordering information. This is often put right on the engineering drawing so that the materials can be referenced by number with the parts. A separate copy of the bill of materials is used for cost estimating, which consists largely of calling up suppliers to get prices. One small hint: when asking for a cost, always ask for a budget price, never a quote. Also ask for an OEM (original equipment manufacturer's) budget price, which carries the largest discount, since either you or your licensee will be doing the manufacturing. If the suppliers want to know what quantity you need, ask them what their price breaks are. In other words, what volume do you have to buy to get a lower price? Also ask if they would ship monthly against a yearly blanket purchase order.

It's hard, when you've started a small enterprise and haven't bought significant quantities of materials before, to be taken very seriously by suppliers. This is particularly true when you've been buying one-at-a-time parts during the development. Years ago, I asked a purchasing agent to get a gallon of special speckle-finish paint to cover a control box I had made. He contacted several paint stores, which didn't carry it, and then called a large paint wholesaler. He described the paint and the fellow at the wholesaler's said, "I don't have it, but I'll make it up for you. How much do you want?" The agent answered, "A gallon." There was a pause and then a plaintive voice from the phone, "A whole f---ing gallon?"

When figuring costs, remember to include scrap (parts that can't be used), freight in (the charge for delivery to your plant), taxes (get your sales tax exemption number from your state), and any outside processing costs such as pickling, plating, anodizing, polishing, papering, painting, or shearing to size.

Often you will be told that you can't get just what you want or that you will have to take more than you need. The answer to this is to make some more calls. There's always someone who will accommodate you if you can just contact him. If you are doing a great deal of buying, you might find a purchasing agent useful. They can be hired through employment agencies. The best purchasing agents, of course, are those with the best contacts.

There are other materials costs that are usually put in with factory overhead because they cannot be assigned to a given part. These might include electric power, truck fuel, welding gas, cutting oil, cleaning fluids, sweeping powder, fuses, work gloves, and other items. Overhead and direct labor will be discussed in Chapter 8, and Chapter 10 gives guidelines on establishing a selling price for your product. Keep in mind that your selling price will probably have to be double what you are spending on material and labor to make a profit, so do everything you can to cut these costs.

CERTIFICATION AND APPROVALS

Just when you think you're ready to go to market, a whole new set of problems hits you: the world of approvals, certifications, listings, and inspections. These act as a protection for people already in the business and are a real deterrent to small companies and innovators.

Generally, the people involved in certification are fair and honest citizens trying to protect the public. However, as most have been burned before by products they'd accepted that subsequently created hazards, they tend to be extremely cautious; therefore you may have to devote a great deal of time to educating them before they'll move. This is a cost for which most innovators and entrepreneurs never budget properly, and many a new product has not bridged this last hurdle.

A lot of the trouble can be prevented by starting early, long before you are ready to submit a formal application or test item. Like building codes, approvals vary in different states and regional areas. It is wise to write to the state office relating to your product (e.g., health, energy, automotive, building, fire safety), asking what agencies you will need approval from to market in that area. Cities such as New York and Los Angeles have difficult, time-consuming, and expensive approval procedures on many products, so it is better not to plan your initial sales there.

Write the agencies with a brief description of your product and request their applicable standards and application forms. Find out if they do their own tests, as Underwriters' Laboratories and the American Gas Association do, or if not, what testing agency's results they accept. Ask if you can obtain a preliminary review, with the aim of having a staff member point out the potential trouble spots.

In many cases you will have to redesign and redevelop to get the approvals you need. You may have to limit your marketing efforts for a time by accepting restrictions on places of use, types of applications, combinations with other products, operating temperatures, rating classifications, and so on. It can be good strategy to start with a limited approval and then, as everyone gets better acquainted with the product and less concerned about it, work yourself into full approval.

Lawsuits, in which applicants have proved bias or unreasonable procedures, have chastened many approval and standards groups in the past. When I first took my Jet-Heet gas furnace through the American Gas Association (AGA), I was required to submit full detailed manufacturing drawings to all twenty-five members of the approvals committee, including twelve representatives from established furnace manufacturers. In other words, my future competitors were requiring that they be supplied with all my engineering secrets before they would decide whether I would be allowed to compete against them. Actually, they wound up approving the project, but later, when other manufacturers had their new designs for vent dampers turned down, a suit forced AGA to change its procedures. Today, the experience I had could not occur. Members of standards committees now have greater respect for applicants. They also have limited terms of office, which keeps the committee as a whole less biased.

When the final hearings on your invention are held, attend them if at all possible. Confidence in the person behind the product sways votes. Then in later years, when *you* are preparing the standards, remember to keep the way open for the little guy.

CHAPTER 6

CONTRACTING DEVELOPMENT AND MANUFACTURING

The great majority of people with bright ideas are so heavily committed to jobs, careers, education, and/or family support that they cannot afford to commit themselves to the full-time venture their invention requires. It has taken a lot of time and effort to come up with a working prototype, a patent application, and, often, financial support from family or friends, but for some reason the next step seems to be a problem. For example:

1. The market for your invention may be too small to make your effort worthwhile
2. The expertise or equipment needed for development may be too specialized
3. Your invention is not proven enough to license (see Chapter 7)
4. You cannot afford the time
5. Your enthusiasm is beginning to fade

If you intend to be an occasional inventor rather than a professional one, or if you intend your idea to be a means to an end, such as getting you into marketing, which you prefer, then contracting out the development and possibly the manufacturing as well is a wise move.

PRODUCT DEVELOPMENT

As discussed in Chapter 5, development covers a broad range of activities and virtually everyone gets outside help at some point. My company has helped develop about three hundred new products over thirty-five years for other people. The vast majority of our clients have been corporations, large and small, who needed expertise, particularly in the field of heat transfer or fluid flow.

It is necessary to distinguish between the G.C. (general contractor) developer and the subcontractor. My company was usually the early-stage G.C., but we got lots of help from subcontractors. We specialized in hardware rather than fancy reports and came up with a working prototype, which was then more critically designed by the client, so that the product's cost and appearance were what they should be, before he started manufacturing.

If you do not have the skill, facilities, or time to either design or develop your invention yourself, you need a product architect or, as they are usually called, an industrial designer. There are many famous names in industrial design: Raymond Loewy, Henry Dreyfuss, and Walter Dorwin Teague, to name a few large firms, or Charles Butler and Manfred Hegemann, who graduated from large firms and set up their own shops.

The industrial designer is most concerned with appearance, as is an architect, but is also vitally concerned with function, structure, and performance. He knows and supervises the contractors who take care of the subsystems. The overall vision is the key. What need will your product fill and what will the market pay for it?

I remember talking to a designer named Jerry Murray at Loewy's one time about a magnetic food mixer that I was interested in and he was developing. On the next table was a layout of a movable, covered, telescoping boarding and deplaning ramp for airline terminals. I expressed amazement that anyone would consider such an expensive thing when metal stairs on casters were perfectly adequate. Murray told me that such ramps would be used as standard equipment in twenty years. He was right. He had the vision to see what the market would pay for this convenience.

Your other option is to hire an experienced product developer, who concentrates primarily though not necessarily solely on getting the invention to function as it should. You can then either do the final design work yourself or contract a designer to put the finishing touches on the product.

Development is an expensive business. It is particularly expensive for the individual inventor, for whom $10,000 is a lot of money to spend on an idea that may not work and which, if it does work, may not sell well. (To a corporation, this is a more affordable sum.) In 1982 the government and research institutes considered one professional man-year worth $100,000, including the necessary secretarial, travel, and other expense back-up but not including the required laboratory technicians, mechanics, draftsmen, and so on. I could give you the names of institutes and universities that do development work, but you'll be shown the door if you're not prepared to talk in professional man-years.

Appendix D lists a number of well-established companies that will do development at a reasonable cost; it should at least give you something to start with. Here are some of the things these developers say:

- Independent inventors have no idea of development costs.

Contracting Development and Manufacturing

- I would appreciate advice on how to deal with inventors.
- I am an inventor and know what they want.
- We are happy to sign confidentiality agreements.
- We will help in negotiating license agreements.
- Quoting a firm price is an impossible task.
- Our rate is $35 per hour.
- If the inventor would take a little more time and prepare proper blueprints, he'd save a lot of money.

The problems developers have with inventors center around money and communicating exactly what is wanted. A promise of confidentiality and help toward getting a licensee are easy for the developer to give, but finding out just what will satisfy the inventor and giving him a reliable cost estimate for something so intangible is very difficult.

A highly successful surgeon once hired my company to develop his idea for a surgical stapler that could be held in one hand like a surgical clamp. It had a good patent, which we assigned to him, and it entirely fulfilled his specifications. We were astounded to hear, even though it was many years ago, that he thought the $2,300 charge excessive. We thought it was a bargain.

Obviously, model building or assembly is a key ingredient of the development process, but the main expense comes in figuring out what it is that the inventor wants built. If you can find someone who charges $50 per hour but can come up with a good design in twenty hours, it costs you only $1,000. However, if you hire someone at $10 per hour and he has to build five models before he gets a good design, it may cost you $10,000 or more.

Perhaps more than anything else, experience is needed to put a good design together quickly. This means mainly familiarity with similar ideas, remembering what was done before, and knowing who provided the components. An extensive catalog file is an important part of a developer's operation. Ask to see a catalog of a certain type of part to get an idea of what is in his file. It's a good clue to his business stability and experience.

In preparing this book, I did a mail survey of some three hundred model-building and invention development firms whose names were selected from recently compiled lists. Over half the envelopes were returned as undeliverable because the addressees had moved. This indicates to me that, although there are people who want to develop and build models, it's a poor business, partly because of the problems involved in dealing with independent inventors. Let's see how we can try to make it work better for both parties.

THE DEVELOPMENT CONTRACT

When someone is going to do a small job for you, there is usually an agreement on a fixed price or an hourly rate plus materials. Often, the financial arrangement is not even written down. If you retain a lawyer, for example, the hourly charge is frequently not mentioned. If you hire a house painter, on the other hand, you generally get an estimate and perhaps sign it as approved.

An invention design, however, is quite different. First, the end product is an unknown; the development process is somewhat like commissioning a piece of sculpture, except that it's harder to draw a picture of your invention in advance and no one knows whether it will work technically and fit the needs of the market. Second, your idea is an "intellectual property," not just one object, and the point is to capitalize on the intelligence involved. The product must be designed within the limits of patentability. Third, there must be constant communication between the inventor and

Contracting Development and Manufacturing

developer so that the result will achieve the inventor's goal, not the developer's idea of it. The house painter knows when his job is well done. The lawyer knows when he has won his case or drawn his contract reasonably. But the developer always wants to improve his product.

A simple contract is essential. It can be a letter-style agreement which one party writes to the other, describing his understanding of the deal and then asking the other party to sign and return a copy of the letter.

The letter contract should have three parts. The first states the agreement for the development of the invention on a "confidential basis," including the provision that all know-how and inventions generated during the development period become the property of the inventor. The word "confidential" is important legally, as I have said before, since a judge will quickly recognize the type of relationship intended and any violation thereof.

Second, the contract should describe in some detail the goal of the intended work, the technical background, the drawings to be followed, and the specifications. It is also a good idea to point out some of the "shall nots" and "are not to be's," so that the developer will know he is not to follow certain paths. For example, sizes are a question. Although development of a very small model may reduce the costs of model building and testing, it may increase development costs, since certain problems may be present in one size and not in another.

The financial deal should be spelled out in dollars per hour for each category of worker with a "not to exceed" dollar figure set for each month, to make sure that limitations on the level of effort and ability to pay are clearly understood.

Finally, the letter agreement should include the right of the inventor or developer to terminate the contract upon agreed notice and under certain agreed conditions, and, for the developer, a statement that this is a "best efforts" agreement which does not bind him to unreasonable standards of performance.

Below is a sample letter agreement of the type often used. Though not the most thorough document legally, it is clearer than a purchase order and well suited to the vague goals of the development business, where the intent of the parties is primary. After you've drawn up your own agreement, but before it is signed, have your lawyer review it to make sure it covers all the important contingencies.

ABC Development Co.
(Address)
ATT: Mr. Robert Jones, President
RE: Invention Development Agreement
Dear Mr. Jones:

Confirming our conversation on __(date)__ at __(location)__ this letter is to recite our understanding regarding your development work on my invention entitled ____, U.S. patent application serial no. ____, dated ____, which you are to perform for the account of the undersigned.

Information regarding this invention is to be treated by you on a confidential basis and not to be disclosed by you or your employees to third parties without my permission. All know-how, improvements, data, drawings, and inventions relating to this invention generated during the period of this development are to be the property of the undersigned.

The work required is to develop my prototype sound- and heat-sensitive light switch into a highly reliable product that will turn off lights in a room only when no one is occupying it. Our goal is that it is to be designed to fit into a standard wall switch box per attached drawing no. ____, and that it sell for under $100. The infrared sensor is to be sensitive to people but not lights. The sound sensor is to be sensitive to fluctuating noise generated within the room but not to motors or fans or to outdoor noise.

We agree to pay you at the rate of $25 per hour for engineering time and $20 per hour for shop and technician labor. Materials are to be billed at cost plus 50 percent. However, in no case is the monthly charge to exceed $2,000. Unless otherwise agreed by the parties, this work may be terminated on two weeks' notice by either party, provided all charges within the cost limitation are paid and necessary shop and test work under way on a particular model are concluded.

While this agreement would not be made unless it was thought that the specifications would be met, it is understood that there is no guarantee on the developer's part to provide more than his best efforts.

If the preceding accurately describes your understanding of the agreement, please sign and return the enclosed copy of this letter.

Sincerely,
SMITH CONTROL CO.

L. J. Smith
President

AGREED:

_____
    NAME

_____
    DATE

COMMUNICATION

Once the work gets under way, frequent communication between inventor and developer is essential. Handling it so as not to inhibit the creative job to be done is a bit tricky, however. What worked for me was a weekly phone call, preferably initiated by the developer, not the inventor. This has to be part of the agreement so it is not perceived as an added nuisance. This phone call should be frank and should describe the problems, decisions to be made, reasons for delays, and perceptions of how the overall goal is affected.

In addition, meetings should be held once a month (perhaps more often in the beginning) to weigh the progress against the cost. After a review of the problems raised in the phone calls as well as any new ones, there should always be a time to discuss new ideas or improvements. There should be no reluctance to discuss problems; if there were no problems, someone else would have already invented your product.

Similarly, improvements leading to new patents are to be encouraged by an inventor, even though they may not be his idea. When an improvement patent is filed, the primary sponsoring inventor handles the paperwork and fees involved but he must by law list it under the name of the person who had the idea for the improvement. The improvement inventor then assigns the patent to the sponsoring inventor. Improvements are the usual course of events in development, and concern should be more for their absence. Without improvements, adequate thought may not be going into the development. With improvements, a "patent portfolio"—which is much less likely to be challenged legally than a single patent—is built up.

The question of credit for the improvements is in the mind of most inventors. What we forget is that it is *not* in the minds of others. They are going to credit the first person they associated with the general idea, not some subsequent improver. Also, remember that if you hold the new-idea sessions regularly with the developer and his aides, if you are encouraging and offer your own suggestions as to how the new ideas can be perfected or incorporated into your invention, you will become a coinventor of the improvement patent. There is no regulation as to the order in which coinventors' names must be listed, so if your name is Smith and you list yours first, the improvement patent will be known as "Smith et al." Your original patent will remain just "Smith."

On the other hand, listing a coinventor's name first might be a real stimulus to that person, and this might be very good for you in the long run. But it could also backfire, as it once did for me. I listed the name of an engineer who worked for me as the first coinventor on a patent that had been 90 percent my idea. He left our employ soon after having charged materials for his new house to the company and refusing to repay it all. This patent became the subject of a Patent Office interference action and this engineer received considerable attention for work that had little to do with him. Many times during the testimony, I regretted having been so generous.

INVESTMENT IN SPECIAL TOOLS

One of the toughest questions is when to invest in special tooling to make your product look and work better and reduce the cost of repeat parts. The developer may have some special tools, that may save you money, but often you must pay him extra for the tooling that he will use but which will be your property. Consider these questions:

1. Whom are you trying to impress?
2. Does the tool solve a major technical problem?
3. What is the payback on the current run of parts to be made?

It may well be that your partner or financial backer would put up the additional money needed if he saw something attractive that was close to being salable. This is a common and legitimate reason for the investment, because you know that if your associate sees only a continuing mess, there will be no more project, even though you're saving him money by getting that last bug out.

Often you and your developer have many problems licked but several left to conquer. You feel one or two could be solved easily if you had the right tools. If you can manage it financially, it may be a real morale booster to get those easy problems behind you; besides, it may well improve the efficiency of the development work.

Sometimes parts are very costly to make by hand—for example, it is expensive to "hog out" a metal bar to a specific profile on a milling machine. An aluminum extrusion can sometimes be used instead, and a custom-made extrusion die costs surprisingly little, perhaps only about $300 for simple profiles (more involved ones, however, may run up to $2,000 or more). The extrusion is dressy, accurate, low in cost, strong, light, and can easily be finished on its surface by several processes. Stampings can also be good investments, but they usually have higher tooling costs. Metal spinnings for round sheet-metal shapes can also be good.

Assembly jigs—arrangements for holding pieces in correct alignment while work such as fastening or welding is done—can improve a product greatly. Wiring and tubing can be bent neatly with very simple tools. In short, there are many good investments and many others that don't make sense. You must be familiar with all sides of the question before you can make the right decision. If you can't put in that much time, make sure the pros and cons are spelled out to you before the money is spent.

INDUSTRIAL DESIGN

If you picked an industrial designer to be your product architect, as discussed earlier in this chapter, you are in good hands as initial development is completed. If not, consider retaining one now. An industrial designer can open the door to product acceptance in many ways.

Virtually all consumer products and their packaging were created with the help of industrial designers. Innumerable items owe their commercial acceptance, at least in part, to good design. These range from automobiles, hotel lobbies, airplane cabins, telephones, microprocessors, copy machines, and lawnmowers to office furniture, fabrics, lamps, calculators, razors, and ball-point pens.

Industrial designers in your area are listed in the Yellow Pages. When choosing one, ask if they can help you take the product all the way through to production. Field and market testing are sometimes part of the service. Costs are high, but at this point confidence in the product should be growing and a professional can keep it building until the final stage.

Even though you are having all this work done outside, you should get involved in the field and market testing yourself (see pages 98–100). This is where you will find out what you've got and receive the most honest answers. Third parties don't have the same point of view because it's not their money, and only you know what the invention is worth.

SUBCONTRACTING THE MANUFACTURING

Having completed development and tested the product thoroughly, you may now want to license it—the common next step (see Chapter 7). However, if you decide not to license or sell the rights to the invention but want instead to be an entrepreneur, you still have choices. You can manufacture the product yourself, have it manufactured for you, or arrange for a combination of both. Also, you can choose whether or not to market it yourself.

One of the most common practices is for a company to represent itself as a manufacturer but have most or all of the factory work done under contract by others. As is very well known, the largest manufacturers in the automotive, aircraft, and electronics industries, as well as many others, are mainly assemblers. What is important is who takes the responsibility, not who actually performs the work. Do what you are comfortable with and good at. Don't manufacture if you don't really like it.

There is a big difference between manufacturers who sell their own products through other manufacturers in so-called OEM* arrangements and those who make someone else's product for them. Ownership and responsibility are again the key. When you have invented the product, designed it, and hired someone to make it for you, it is your product and you need not have a shop or plant unless there is some good financial reason.

Most important is finding the right company to do the job. In the hundreds of times over the years my company has had trouble with a supplier, we have almost always run into a better supplier we wished we'd known about before. It makes sense. When you first start looking, you are eager to find someone, he is eager for the job, you are susceptible to flattery, and the guidelines are vague. It's a trusting arrangement anyway, so you take a chance on one of the first candidates. But as soon as you start to buy, word gets around and other, often better suppliers contact you.

Of all times in your business, now is the time to be

---

* Original equipment manufacturer; see page 146.

choosy. Be sure and talk with other customers of the manufacturing company. Follow up phone calls to references with a personal visit to find out what advice they may have for you. So much depends on this; you must get all the help you can.

The company's financial status is second only to the references as a criterion for selection. When I was a subcontractor making Roll-A-Grill hot dog cookers for the fast food industry, Jack Connolly, a customer with a sage Irish wit, used to say, "Always do business with people who have money; it's surprising how much rubs off on you, and vice versa."

Taking the lowest bidder is not a good idea, although it is good to get a number of bids. The wise subcontractor will jack up his price against unknowns until you have worked together for a while.

Deciding how much of the manufacturing should be done outside, or whether you want to reserve something like assembly for yourself, depends partially on your other commitments. It is better to err on the side of having more done by others, particularly if you are involved in marketing. The fewer new jobs you have to undertake at once, the less risk. This is assuming, of course, that you've selected a good manufacturer. In a small business, an oft-repeated phrase is "I should have done it myself."

The investment in tools and dies is easy to determine in some cases and hard in others. Most experienced subcontractors will have strong ideas about this, which you may be obliged to follow if you select them. Even though you are charged for the tools and dies, you may not be able to take them if the two of you part company. Maintenance put in by the subcontractor may be figured as an investment, and thus possession may be retained. This must be agreed upon beforehand. Basically, to remain in business today, tooling is essential and must be done at the outset. It should be figured in all your budgeting and fundraising. Investors are more lenient toward this expenditure than most others because they know the potential return.

If time is very important to you—and it usually is, since money has a way of running out—consider a bonus/penalty clause for delivery schedules. If the subcontractor beats his schedule, he gets a bonus; if he's late, he gets a penalty. Usually, the penalty is worked out on a daily basis: you will get a certain amount of credit on your bill for each day the subcontractor is late.

The main problem with this is that if the subcontractor hurries, quality control and testing are shortchanged. The test standards and drawing tolerances thus need to be gone over very carefully by everyone involved. A good practice is to mark in yellow pencil on the drawings which tolerances are the key ones that you will insist on. Give the subcontractor ample prior information.

Tests should be devised that can be done quickly and simply with little chance of error. New digital readout meters, for example, speed up reading and reduce error; using the old instruments with scale readings is wasteful. BTU meters combine flow and temperature differences into one. Of course, the minicomputer field has endless applications for testing controls that are being developed rapidly. But novel ways of getting at what's really important are much more valuable than sheets of worthless printouts. Large volumes of computerized test data are never used and often never looked at.

Tests should be run by someone who is not involved in making the product, but has a good technical background. Engineers in training or laboratory technicians taking evening courses can often be found for this work, since it is usually part-time. The production manager should not be in charge of the test personnel if at all possible. The temptation to pass marginal results is too strong. On the other hand, a tie-in with engineering is good because your testers will be looking for causes of problems and ways of correcting them at the same time.

THE SUBCONTRACTING AGREEMENT

The nature of a subcontract agreement depends on the parties involved and the profitability of the product. Its success depends in part on communication. The best thing you can do after production is under way is to keep the channels of communication active by visiting the plant and sending memos that show you are interested but not complaining. Be complimentary and appreciative of the work whenever possible.

The contract is more than just a purchase order from you to your subcontractor, for it should spell out the entire relationship. Again, confidentiality is important. Your proprietary position must be emphasized. Warranty provisions; handling of returned goods; drop-shipping to customers; ownership of tools, dies, drawings, specifications, prototypes, and improvement ideas; terms of payment; testing; quality control; inspection of records; authorization of changes; duration of quoted price; method of packing and shipment; commitments as to quantity and delivery; handling of sales taxes; spare parts; overtime; limitations of liability of both parties; and many other points are best settled in the beginning or before the first production run.

The following agreement, with the names changed, has held up successfully for twenty years. It takes into account most of the items recommended above, but in the actual operation there are always compromises to be made when problems arise. It is reprinted here to illustrate the basic form these contracts take. When you are drawing up your own agreement, you should have it reviewed by a lawyer before signing it.

Mr. James R. Brown, President
Ajax Mfg. Corp.
500 First Avenue
Chicago, IL 50103

DATE: _____
RE: Supply Contract for Floral Power Units

Dear Jim:

This agreement replaces the present temporary agreement between us covering your supplying us with Floral Power Units per our Drawing No. C-1001 except that present price agreements shall remain in effect subject to increase, not more often than once a year, upon evidence of increase in manufacturing cost.

It is agreed between the signatory parties that:

1. Floral Power Light Co. (hereinafter called "Floral") agrees to order all of its requirements for the above product exclusively from Ajax Mfg. Corp. (hereinafter called "Ajax"), and Ajax agrees to supply all of the above products exclusively as ordered from Floral and to ship them only as instructed by Floral. However, if Floral shall receive a comparable bona fide quotation verified by Ajax more than 5 percent less than the price then in effect by Ajax, Floral may change its source of supply on such item unless Ajax shall reduce its price to within 3 percent of such quotation, and provided that Floral's accounts invoiced and shipped by Ajax on such item are paid in full, and provided further that all finished goods of such item shall be bought and paid in full and all raw material, work in process,

and irrevocable orders for material shall be acquired and paid for by the new supplier simultaneously. Floral may also change its source of supply if Ajax is prevented from shipping for over thirty days because of bankruptcy, insolvency, or act of God provided the above provisions for payment for all accounts, finished goods, raw material, work in process, and irrevocable orders are consummated.

2. Terms of payment are 2 percent ten days, net thirty. Ajax shall not be required to ship if payment terms are not met.

3. Ajax agrees that these products are the proprietary property of Floral, that all improvements and patents are to be assigned to Floral, and that Ajax shall not manufacture such product for anyone else for a period of ten years after Floral changes its source of supply of such item as above, provided that Floral continues to offer such product for sale.

4. Ajax agrees that Floral has paid for and owns the tools, dies, and fixtures used in the manufacture of these products with the exception of those made by Ajax for which Floral has not been invoiced. Ajax agrees that upon Floral changing its source of supply, as in Paragraph 1 above, Ajax will immediately sell and deliver to Floral all such tools, dies, and fixtures at their current value based on original cost and estimated life and will also immediately deliver, insofar as it has the power, all those owned by Floral. Floral will be entitled to seek immediate injunctive relief to gain possession of these items it owns.

5. Ajax pledges to secure agreements from its key personnel to the effect that they will not use the knowledge of any of Floral's products gained during their employment with Ajax to manufacture or

market any of these products or similar products, either alone or in conjunction with third parties, or in any way to advise, counsel, or assist any third parties to compete with Floral.

6. Ajax agrees to use its best efforts to secure high quality and workmanship. Ajax warrants all assembled units and parts sold to Floral under this agreement for a period of thirteen months from date of shipment against defects due to faulty workmanship or materials. Ajax's liability hereunder shall be limited to the obligation to replace without charge defective units or parts thereof returned to Ajax and found to be defective as aforesaid. This warranty is expressly in lieu of any other warranties, expressed or implied, and shall be voided by any tampering, repairs, or replacements made other than at Ajax's factory or by its authorized representatives.

7. Floral agrees to use its best efforts to sell and promote all of the above products and to keep Ajax closely advised of its estimate of future sales.

8. Floral agrees to hold Ajax harmless from any patent or trademark infringement.

9. If Floral ceases ordering any or all of the products for a period of 90 days without arranging to purchase the inventory, Ajax may make such arrangements as it wishes to dispose of the inventory including selling finished goods in the market. Ajax may terminate this agreement upon ninety days' written notice, and Floral has the option to acquire the entire inventory over the following six-month period; if it does not acquire the entire inventory, Ajax may dispose of such inventory in any way it chooses, including selling finished goods on the market. Thereafter, Ajax may not manufacture or

sell such goods for a period of ten years from the date of cancellation.

10. This agreement shall not be assignable by either party without permission of the other party; but if such agreement is given, the assignee shall be bound by the terms and provisions hereof.

If the foregoing is in accordance with your understanding of our agreement, please execute both copies of this letter and return one copy to us to signify your acceptance.

Sincerely,
FLORAL POWER &
LIGHT CO.

BY: _____
John Smith
President

ACCEPTED FOR:
AJAX MANUFACTURING CORP.

BY: _____
    James R. Brown
    President

CHAPTER 7

LICENSING OR SALE

The ideal answer for an inventor is to license his invention to an established company that can manufacture and market the product. Unfortunately, this is not as easy as it might seem. Problems of legality, economics, timing, and communications often destroy what should be a good thing for all concerned.

FINDING A LICENSEE

When you write to a company asking whether they'd be interested in your idea, you automatically get back a nonconfidentiality agreement saying that they cannot consider your idea unless you agree beforehand to two things: that no confidential relationship is being established and that you will rely on your patent or potential patent protection for your legal rights. This seems unfriendly and offends many inventors; however, there is a good reason for it. The company may already have been doing development in the same field and they don't want that work limited by anything more than the specific patent rights to which you are entitled.

You can tell a good deal from the way a company handles its reply. Many companies feel that anything that did not originate from their own employees could never be as good, and they do not want your ideas. If their NIH ("not invented here") factor is high, their form letter will be blunt, legalistic, and a turnoff.

If a company really wants ideas submitted, you will know. Again, it's better to go in person or at least telephone to sound them out. Often companies go through cycles when they are just too busy to consider anything new. When the U.S. Department of Energy instituted efficiency standards for most types of appliances and home energy devices, many companies' engineers were so tied up that they were fully occupied for more than two years with tests on their existing products. Innovation was totally blocked. A visit would certainly uncover such a situation, and a phone call would too, if you ask appropriate questions.

In selecting a licensee, it is often wise to pick a company that would like to get into your product's general field but is not yet involved in it. When I had invented a blanket that worked by circulating warm water in tubes the size of wires, I made the mistake of licensing it to Fieldcrest Mills, who were number one in electric blankets, selling over 2 million blankets a year. They told me their electric blanket business wasn't as profitable as they would like it to be and that the more even comfort of the "water blanket" would bring a higher price and give them something new over the competition.

Unfortunately, the water blanket's biggest selling point was that electric blankets cause fires and that a large number of people are afraid of having electric wires covering them at night. Fieldcrest would never say that for fear of hurting their major business. Furthermore, the organization, all the way down through the wholesalers and retail outlets, were trained in and loyal to electric blankets and would never give water blankets a chance. Instead, salesmen called it a "wet blanket" and never pushed its safety or comfort features. That product reached a volume of only 3,000 per year and was dropped in three years because of the $25,000 per year minimum royalty. By that time other manufacturers concluded that if Fieldcrest couldn't sell it, no one could. It was never licensed again, but my wife still thinks it's my best invention. It would have been better to take it to a towel or appliance manufacturer that did not have a big stake in electric blanket business already.

In contrast, licensing professionals like Robert Goldschieder of the International Licensing Network, Ltd., in New York City go to great lengths to find a potential licensee whose business will benefit in every way from the license and who will make it part of the company's long-range planning. Here is Goldscheider's plan as he gave it to me in a recent interview:

1. Search out a licensee in terms of its ability to sell, not to engineer or to manufacture. (For example, stick to those that advertise heavily.)
2. Avoid internal conflicts of interest. (For example, if your product is a plastic chain, avoid chain manufacturers and find plastics marketers.)
3. Get through to a decision maker. (Call the president's office at 9 a.m., and someone will put you in touch with the person who can give the project the presidential seal of approval.)
4. Prepare a licensing memo of seven or eight pages containing:
   - The history of the innovation
   - Background on the inventors (their qualifications)
   - A rundown of the market and economics of the invention

Licensing or Sale

- The package of intellectual property (patent, trademark, lawyer's opinion and reputation, etc.)
- What deal you want (for example, a three-month option for $20,000 with right to renew at the same fee, with half of option fees credited against final license down payment—see the section marked "option" on page 132).

5. Try to connect your invention to a scholarly article.
6. Find out what is a must for the licensee's side, and agree to it only after many concessions on their part.
7. Make yourself look successful. (Pay attention to shoes, fingernails, clothes.)
8. Rehearse negotiations, getting someone to play the role of prospective licensee.
9. Offer five days free consulting to give the licensee time to get the reaction of his key thinkers).
10. Use *your* license form for discussion, not theirs.

Appendix E ("Some Companies Looking for Inventions") lists a number of companies in various fields that are actively looking for inventions. I hope that this will give you a start in finding the right one for your idea.

If you do not want to do the licensing yourself, maybe you will be lucky, like Dr. Fritz Wankel, who in 1960 arranged for Curtiss-Wright, the New Jersey aircraft engine maker, to be the master licenser on his rotary engine. In this way, he licensed Ford, General Motors, and Mazda for huge sums.

SHOULD YOU LICENSE, OR SELL?

An alternative to licensing is to sell your patent. This is purely an economic transaction based on the following considerations:

- How valuable is it now compared to what it might be worth later?
- How long a period does the market have?
- How badly do you need the money now?
- How eager is the other party to own it?

Licenses are risky, require constant attention, can be cancelled unexpectedly; they often involve lawsuits and may be upstaged by other new developments. On the other hand, the product's value increases with the success of the license and with inflation, and a license income can grow over a twenty-year period to many times the original sale value. Attractive sales offers are rare, and taxes can take a bite out of a sale.

There are, however, a number of good reasons to consider selling rather than licensing your invention. If you've brought the product to market and it is doing very well but you doubt it will last more than a few years, sell it. If you desperately need money to avoid bankruptcy, sell it. If your customer wants it very much to round out a line of his own products and portfolio of patents, sell it.

Keep in mind that if you push for a sale, the other party may fear that you know of or suspect some serious, hidden problem and that you want to get your money and run. Let the other party ask for a sale; you can push him into it by setting your down payment request very high.

NEGOTIATING THE LICENSE AGREEMENT

Once you have an interested prospective licensee, how do you proceed? Here are some principles to keep in mind during negotiations.

Initial Payment

Make it clear from the beginning that there must be a down payment upon signing. They will argue that they must make a big investment in design, further development, tooling, factory setup, and so on. However, if they have made an initial financial commitment, they will be much more inclined to invest in design, development, and so forth. Your invention will have a much higher priority for them. If they won't make a down payment, the deal would not be worth much anyway.

Exclusivity

An independent inventor rarely has a product that merits multiple licensees in the same geographical and product areas—though it does happen, as in the case of Dr. Wankel. Exclusivity is generally the expected thing, and it is better for you. Sublicenses can be granted if warranted, and both you and your licensee will share in these.

The Grant

Regarding the grant, permitted uses, territories, and time period you are granting the licensee, try to limit them to what will actually be pursued. If you grant foreign rights, you will probably be obliged to maintain foreign patents. Take it from someone who has spent a lot on them, they don't pay. If your licensee insists, let him pay to file and maintain them and drop the royalty rate. Canada is the exception; it charges no yearly fees, involves no language translation cost, and is close enough in both distance and type of market to be a good investment.

Royalties

Royalty rates can vary from 1 to 10 percent depending on circumstances. Most often, for a mass-produced item under exclusive contract, 3 to 5 percent of factory receipts is reasonable, perhaps declining somewhat with quantity in order to provide more promotion funds. Improvements made by the licensee's employees are a sticky item. If possible, try to get title to them as part of the license agreement; argue that they wouldn't have gotten into this at all if not for you, and that if a division of interest is set up, it will impose a wall of secrecy that will impede the cooperative effort needed for success.

Minimum Royalties

Minimum periodic payments creditable against earned royalties are usual in order to prevent a licensee from shelving a project. However, if this cannot be worked out with amounts large enough to be significant, there is a fallback position to minimum performance: the licensee guarantees to spend an agreed minimum amount of money on salaries and expenses for the project or forfeit the rights. This brings you no money, but the expenditures are intended to be a spur to the licensee, not a reward to you. This requirement can be divided up among several different uses, so that if your patent has two or more product applications, you could get some of them back if they are not pursued. The dollar figure must be high enough so that the licensee won't put your product on the shelf just to keep it off the market. A "best efforts" clause also helps with this (see page 110).

Warranties

There are a number of provisions you should include even though the licensee may protest that they are meaningless: a "best efforts" clause, permission to examine accounting records, agreement to mark product and promotional literature with your patent number, interchange of technical and market information, and agreement to settle differences by arbitration.

A licensor usually cooperates by agreeing to add product improvements to the license without charge, supplying all the information the licensee wants, permitting sublicenses, and allowing assignment of the license to a successor or subsidiary.

Your Continued Involvement

The chances of a license working out for you are much better if you stay closely involved without being a nuisance. If you can get a consulting contract, be a supplier, or become a customer in a limited way, you will have a reason to stay close. A consulting contract is just a provision in the license agreement for your presence on a certain number of days, or days per month for a given time, for expenses only or for expenses plus a stated daily fee. The licensee's employees, who might resent somewhat having to work on your invention, will be much more willing to operate under a royalty agreement if they know and like you. Also, you will be aware of problems that arise and have a chance to help correct them before they become serious. Thus, you can at least be a coinventor of any technical fix. The whole project is a joint effort and you should be as supportive and noncritical as possible.

Infringement

Defense of infringement charges from third parties are your responsibility, but you should try to limit your liability to the total amount you have received from the licensee. You will rarely face a suit without a chance to "cease and desist," so just be sure that you have no liability to the licensee in case they are forced to change design or even quit entirely.

Termination

Termination of the license by the licensee to avoid minimum royalty payments is often set for one day of the year, with at least ninety days' prior written notice. This allows problems to cool and be settled without threats and assures you of another year once the notice date is past. You will be able to cancel only upon breach of contract, such as not being paid what you are due, and then only after the licensee has been given a chance to correct the deficiency.

The Use of Lawyers

The licensee will probably have his lawyer in on the negotiations, but it is not essential that you do, and it is expensive. The business deal is for you and your business adviser to negotiate. But it is wise to say as a fallback "I have to check with my lawyer" and consult with him or her later. Your lawyer can look over a draft agreement prepared by you or the licensee's attorney and make any changes felt to be necessary.

A typical license agreement is reprinted in Appendix F. It has proved quite successful when used by a large company as licensee and a very small company as inventor/licensor.

THE OPTION

A potential licensee may not be ready to make a final decision for one of many reasons. Yet they might be willing to take the time and effort needed to come to a decision eventually if they know the deal will be available for them. In a case like this the option is a commonly used alternative. You sell an option to the licensee, giving them the right to take up that license—for a price—within a certain period of time. The option fee may be wholly or partially creditable against the license down payment. Obviously, it means negotiating two documents, the option agreement and the full license. A lot of business is done this way, as in the publishing industry, where books are optioned for the movies or television. Don't count on the option being exercised, though. It is wise to find another interested party just in case.

A sample option agreement is reprinted in Appendix F, page 188.

HOW WILL YOUR LICENSE DO?

Now that you are one of the very select group of inventors who have licensed their product, what will happen to you and your licensee?

Statistics show that very few inventors can support themselves on royalties alone. The average income is well below the poverty level. Thus, you must plan to support yourself in another way and use the royalty income as a bonus. On the other hand, chances are about even the royalties will increase and you'll eventually do well.

A study by the management consulting firm of Booz-Allen & Hamilton, as condensed in *Invention Management*, reveals that:

- Two-thirds of the new products introduced between 1976 and 1981 by the 700 major companies surveyed were successful and three-fourths would probably stay on the

market. However, half of these were modifications of existing products.
- Only 10 percent of the products were "new to the world," but 27 percent of the big winners were "new to the world."
- One out of seven of the ideas seriously evaluated by the companies turned out to be a successful product, compared with the ratio of 1 in 58 that prevailed in an earlier five-year study. (One reason for the improvement is that weak or unsuitable ideas are now weeded out much earlier.)
- Successful new-product companies don't spend more on R&D than unsuccessful ones, and there's no correlation between the number of new products and success rates.
- By far the most important reason companies take on any new product is technological advancement.
- By far the biggest internal obstacle to making a new product successful is current business pressures on top management.
- Companies expect to double the number of new products they bring out.*

These findings, by and large, are more favorable than ever before. In a nutshell, they indicate that although you started out at 1 chance in 100 for a successfully marketed invention, your chances improve to 1 in 7 when you get a reputable company to make a serious evaluation. If you conclude a license, your chances have drastically improved again to 2 out of 3. Companies have learned a lot about how to handle new products. Our economy badly needs new products. There was never a better time for inventors.

Invention Management, vol. 7, no. 2 (February 1982), pp. 1–2.

CHAPTER 8

SETTING UP YOUR OWN PLANT

Just because you're an inventor doesn't mean you have to be an entrepreneur as well, manufacturing your own invention. You can, as discussed in Chapter 6, contract the manufacturing out. But if the EP bug has bitten you, you should be aware that setting up your own plant can absorb endless amounts of time, energy, and money—if you let it. Or it can be a most rewarding experience, providing you with a stable future if you're lucky and can control the business's demands.

To me, a plant has been invaluable. When I left GE thirty-seven years ago to start my own manufacturing business, the chief concern of my friends was job security, the high risk of failure, and fear of being out of work. What amazes me is that although I've been through some hard times, I've had one business all these years; the worriers have changed jobs and locations many times, been unemployed for brief periods, and are now being forced to retire at age sixty-five, live on small pensions, and wonder whether they can start businesses of their own so late in life. I'm still trying to raise money to expand into new ventures, but at least I've had a lot of experience in the meantime.

The best rule is to start small, or "creep before you walk." If you can find the right opportunity to buy out an existing business, you'll be way ahead. Place a classified ad in your local paper, under "Business Opportunities" or "Business Wanted," in which you offer to take over and run a small business plant with machinery, materials, or products similar to those involved in your invention. Offer to pay most of the purchase price over a period of years, since often the seller is looking for long-term income.

You can make your manufacturing business more profitable by doing subcontracted work for others. You will then have the personnel for model building and testing your own future inventions, and you will get a different outlook on development than that shaped by laboratory-dominated test procedures.

One young man who did it right was Murray Green, sales manager for an eastern manufacturer of oil-fired water heaters. Murray wanted to have his own company, felt he knew how to improve oil models greatly, and went looking for a small water-heater manufacturer that was not doing well. He found one making stone-lined electric models, determined that the management was interested in a sale, and lined up a financial backer. He planned to run the company as it was for a year and then move to a more efficient plant and add the oil-fired line. It worked out well, and he soon had his own profitable firm for a fraction of the time and expense of starting from scratch. What's more, he had the facility for new developments that have now put his Ford Products Corporation into the forefront among companies making solar water heaters.

PERSONNEL

Hiring the right people is a tough job. Be even more thorough in interviews and checking references than you might think is necessary; the temptation is to hire someone you like, without really finding out how good he or she is for your work. For a small company, you need unstructured, all-around, self-motivated people quite different from those who do well in the structure of a large corporation.

The kind of person you need likes to know what is going on. The greatest motivator is simply to keep everyone aware of what's happening in the company. Circulate a file folder every week to all office and lab personnel, with copies of every letter, memo, and purchase order sent out during the preceding week.

Employee benefits are much less important in a small company than in a large one. Being in on everything is the big benefit. Pension plans are not worth the trouble in most cases. Long vacation periods and numerous holidays and personal days will not be terribly important to your employees, because they'll feel they're an essential part of the work. Company-paid Blue Cross/Blue Shield, two weeks' vacation after a year, and contributory life insurance are all that's really expected. If an employee is sick or has a family problem, having a generous and understanding boss is far more appreciated than having a flat policy that may cost the boss more.

MOTIVATING WORKERS

In a very small plant, two key things make all the difference in the workers' performance: that they know what's expected of them and that they know it's appreciated when it's done. As soon as possible, set an expected hourly or daily output for each job so the worker can judge how he or she is doing. Post a comparison sheet stating both the output performance for each day and the expected amount. (Piecework, that is, paying a worker for each piece produced instead of each hour on the job, is seldom applicable to new products and is hard to administer.) Also post a simple list of miscellaneous jobs—such as cleanup, maintenance, painting, sorting, counting, inventory, and so on—that workers can turn to after they have finished their regular assignments. You will have the best possible working relationship with your employees if duties and expected levels of performance are spelled out and effort is recognized and rewarded.

EQUIPMENT

In order to justify the investment and the space it will take up, any machine you are thinking of buying, whether for the plant or the office, should be one that will be used a good deal. You will need machines, but keep your investment in them low to begin with. You can always invest more in them as your profits increase. Also, new types of machines will become available, making the old ones obsolete. To justify its cost, a machine bought new will have to see a good deal more use than a used one.

There is probably no market more susceptible to dickering on prices than the used-machinery market. Downtown New York, for example, is an excellent place to go if you know what you need and can afford to spend. Just keep going back and forth between dealers and surprisingly good buys will appear. Auctions, which are often advertised in local papers, are also an excellent source of bargains.

Remember that in buying machinery manufacturers go by the three-year payback rule. The machine must pay for itself within three years—and when you're just beginning, two years is a better time frame.

Machines and other equipment are also used in the office and are just as important to your profit as is the equipment for the shop. The more you get yourself tied up in paperwork, the less you can do about getting orders or speeding up production. A good copying machine, self-correcting typewriter, postage meter and push-button phones, metal file cabinet, customer and supplier bookkeeping files, and well-designed letterhead stationery are all most helpful to efficient operation.

The small office computer has been a real boon to fledgling companies getting started in the '80s. These machines are simple enough that almost anyone can operate them. They can do such a variety of jobs that they benefit almost any operation. If you have personnel bright enough to invent, develop, manufacture, and market an innovative product, they can certainly adapt to this versatile new tool.

It's easier to raise money for a business that investors can be proud of, and a tiny company can look big on the minicomputer's word processor. It gives you a touch of class: letters are professional-looking, what were duplicates can now be originals, corrections are easily made. Accounting can be quicker, cheaper, and much more usable. Forecasting can be worked out with many different sets of assumptions. Technical data sheets can be supplied to sales personnel. Polished-looking reports and proposals can be issued. Financial statements will aid in getting support from banks. Many very small companies are now using this technology; not doing so puts you at a disadvantage. New equipment is coming fast at lower prices, so stay alert. If you want others to use your inventions, you should use theirs.

SUPPLIERS

Everyone knows that customers are vital to any business. Less well known but equally true is the importance of suppliers (also called vendors). A supplier who lets you take forty-five days instead of fifteen to pay him (or instead of requiring cash on delivery, as many do) is as valuable as a customer who pays in fifteen days instead of the average forty-five. Credit is available as long as there is cash flow and your suppliers know you and your business.

As noted before, you will find directories of suppliers in your industry at trade shows, in trade publications, in the *Thomas Register*, regional publishing guides, the Yellow Pages, and so on. Your most important tool in your search is the telephone, and you will spend many hours following leads. A good maxim to keep in mind is that there's always a better vendor that you haven't contacted yet; but when you finally pick one, be loyal.

Visit your suppliers early in the relationship so you know their people and their facilities. They, of course, will be very interested in your success and in not losing your business to their competitors, so they will be eager to help you with technical, production, or marketing advice whenever they can. They'll be able to offer the kind of practical advice you can't get elsewhere, telling you, for example, which vendor of a specific component is reliable and which is not, who is a good distributor in your field, which engineer knows most about a given subject, and what is the trend in materials and pricing.

THE BREAD-AND-BUTTER LINE

Every new plant needs a bread-and-butter line—one that is steady, uneventful, and reliable—to keep it going. Murray Green had his electric water heaters. I had my Roll-A-Grill hot dog cookers. You must have something too before you start out.

To find the product that's right for you, visit local manufacturers and talk to the subcontracting executive at each one or, preferably, the president of the company. Tell him you'd like to take over the manufacturing or assembly of some product he's now making, saving him money and space and letting him concentrate on marketing. Most companies have low-volume or seasonal products that they consider to be a nuisance. Don't take on something new; you have already done that with your own invention. What you want is a product that is low-profit but pays the overhead and takes little of your time. One person's nuisance is another's bread and butter.

Talk to the chamber of commerce, the banks, local suppliers, utilities, truckers, trash removers, the newspaper, local lawyers, UPS, the post office, government employees, and the police. They hear about the troubles of local plants and know who needs help. Trade shows are another excellent source of information and the one that has worked best for me.

CUSTOMER PAYMENTS

How to handle your customers' payments is one of the toughest questions when you are starting a plant. Should you give them credit, COD, require a deposit in advance, or use a different approach? One method that often works for the small, noncompetitive product line is to give a 5 percent discount for an order accompanied by a check. This cuts down on accounts receivable, credit and collections, and interest on working capital. Like any other expense, it can be part of the pricing procedure. We've done it for years, and our customers like it.

MONEY MANAGEMENT

The world likes to think of inventors as poor businessmen who don't know how to manage money. Most of the inventors I know are pretty good at it, and they know how to make what they've got go a long way. They eat brown-bag lunches and drive eight-year-old cars. Almost everything they buy is a bargain. They don't lose money on investments because the money they have goes into their inventions. Their money problems come from risking too much on a difficult invention that took longer to perfect than they figured and from not charging enough for their services or licenses when they do succeed. These are not really money-management problems; they are more a matter of judgment.

Many things are involved in keeping your business's cash flow running smoothly. As this is really beyond the scope of this book, you are urged to consult some of the guides listed in Appendix H ("Recommended Reading"). However, I do have a few tips that might make things easier for you:

1. Be persistent at collecting money owed you when due and charge 1½ percent per month if your debtor goes over the terms you granted.
2. Be stingy about granting credit. Make a rule not to give credit to anyone until they have reordered once or twice.
3. Make up a list of certain of your suppliers as references, but only after sitting down with them and negotiating both special terms and permission to use them as references. Such a list is invaluable to you in getting credit from other suppliers. Keep your payments to your list of references strictly within the terms of your agreement.
4. Be sure to initial every purchase order yourself, even if you get to the point where you've got several million dollars in annual sales.
5. Institute thorough and immediate inspection of incoming materials. It takes little time and can save a lot of expense when problems are caught quickly.

Setting Up Your Own Plant

6. Use single-entry bookkeeping (a "purchased one-write system"), which avoids making three copies, to save time.
7. Pay your employees by check and make whatever arrangements may be necessary for a bank to cash their checks.
8. Treat your business bank account as conservatively as you would your personal account.
9. Be very leery of giving personal guarantees to suppliers or lenders.

A disorderly company rarely if ever makes it. However, the major factors in the success of an invention are the worth of the product and its salability, not the day-to-day matters of keeping things running. Make sure everyone involved in the company understands the invention itself. Don't let your advisers say, "Oh, I don't know anything about technical matters, so you'll have to handle that question." Get them involved in the technical matters. Your marketers and customers would love to have you explain it to them. It's mainly a matter of translation.

FINANCIAL DIFFICULTIES

If you owe your suppliers money and can't pay them on time, phone or write to them and explain the situation. The worst thing to do is keep silent, making them call you and then being evasive. If they know you're having trouble but are doing the best you can and treating everyone fairly, suppliers are usually very cooperative. They will treat you the way they would want to be treated if they were in that position, which most of them will be at one time or another.

And while we're about it, if you think you have to go into Chapter 11 of the Federal Bankruptcy Act, work with your suppliers yourself. Don't get a high-priced lawyer, as I once did when I was in that situation, for you'll wind up doing all the work yourself anyway. Call your suppliers to a meeting and work out the settlement there; let them get the money the lawyer would get in excess of his regular hourly charge for the necessary filings. Tell them that if they can help you get back on your feet, you will buy exclusively from them in the future. Whatever they lose now, they'll get back later.

If you're lucky, you can avoid Chapter 11 altogether, although it takes only one dissident to upset the boat. In my case, my costs for my lawyer, the creditor's lawyer, my auditors, the court auditors, the bank's attorney, the receiver, court fees, and twenty-eight fifty-mile round trips to the Federal Court House would just about have paid the suppliers. In addition, the stigma hurts your business and makes it hard to collect what your customers owe you; they figure you won't be around to back up your product, so why pay? Then you have the cost of suing them plus the inevitable settlement losses. Use a local lawyer who personally knows your bank, your landlord of the last three months, and the local IRS office. These are your three secured creditors who must be paid in full up front. Remember that your equity investors are the lowest in priority and get nothing except maybe patent ownership if you don't continue. It follows, then, that they can be helpful allies in negotiations.

There are many important considerations in running a plant, from handling unions to taking care of such elements of factory and office overhead as rent, utilities, insurance, salaries, overtime, benefits, maintenance, and postage. Entrepreneurs deal with these issues in various ways, and they debate them hotly whenever they get together. Get as much advice as you can from books and other resource material (see Appendix H, "Recommended Reading"), and quiz others in your position. Remember that small companies, particularly start-ups, can get by with much lower overhead than can big businesses. You will find yourself wearing a number of different hats, making the kind of quick decisions that are so difficult to effect in large firms. This is the essence of low overhead and successful small-business management.

CHAPTER 9

MARKETING BY AN OUTSIDE FIRM

Although a great many inventors are highly effective salesmen of their own products, they are rarely also good marketers, interested in designing and carrying out the strategies of effective and expanding product turnover. Inventors are primarily interested in ideas, while marketers want to influence people. Selling your idea and selling your continuing product line are two very different things.

If you prefer not to do your own marketing, you can hire a marketing expert or take in a marketing partner. There are, however, other options, which involve making the product yourself and letting another company handle the sales either in your name or theirs.

ORIGINAL EQUIPMENT MANUFACTURERS

An original equipment manufacturer (OEM) is a company that manufactures and sells its own products but also

Marketing by an Outside Firm

buys assembled products designed and made by others to resell, often under its own label. The OEM assumes all the sales and marketing costs and much of the responsibility. He takes on the mantle of manufacturer and, as far as the world knows, he is.

The OEM buys from the factory in quantity at a lower price than the manufacturer would give to a wholesaler, dealer, or direct-mail customer. Often the purchased product will be integrated into his own; for example, while a tire might become part of an automobile but still be sold under the supplier's name, an air conditioner as part of that same car might be sold under the OEM's name.

Back in the days when the safety razor became highly competitive, the twenty-five advertised brands of blades were made in only four factories. Today, one manufacturer I know makes furnaces that are sold under twenty different labels. It wouldn't be economical for every marketer who must have a full line to supply to his dealers to make all his products himself.

The advantage of selling OEM is obvious: you may be able to establish volume fast with very low sales cost. The disadvantages are also pretty clear: you lose control and you are vulnerable to the whims of the OEM.

The large retailers—such as Sears, Montgomery Ward, and J. C. Penney—who act much like OEMs, are dependent on their suppliers to such an extent that they have strong policies to protect them in the areas of pricing, termination, deliveries, design, and so on. A typically fair Sears pricing policy used to be the cost of materials, labor, factory overhead, plus 15 percent, plus engineering and tooling. Companies who are predominantly manufacturers rather than retailers and are buying outside to round out their lines are more risky because when they need work they'll do their own manufacturing rather than buying outside. In any case, it is wise not to let your OEM sales rise above 50 percent of your total, to avoid being at the mercy of the OEM or the retailer. In other words, sell both OEM and under your own name if you can.

A lesson in how not to handle an OEM arrangement concerns the first portable tape player, called Playtape, invented about 1968. The product was designed, developed, and privately financed; market research was done to show that teenagers would buy an inexpensive portable tape player to have music in their car or at the beach. A small manufacturing space in New York was secured and production begun.

Sears was approached and became interested. Their studies showed real sales potential if the price was low, and they liked the idea of being first on the market. They priced the machine to sell at $9.95 and sent in an initial $2 million order. The manufacturer was overjoyed, rented a factory in New Jersey, bought tooling to reduce the cost, hired employees to meet the challenge, and delivered the goods on time.

Sears never reordered. Different products combined with radios and other features came along. The manufacturer had very few additional sales and high overhead. He spent his remaining funds trying to push sales high enough to break even. His pricing had been set too low and was hard to raise. In spite of all the good work, the manufacturer went under.

Perhaps if the manufacturer had shunned the Sears order and had built up his own sales first at a higher price, he could have handled an order from Sears and brought along new designs to stay competitive. But it's very easy to be seduced by a big marketing name. I was.

Exxon (then Esso) and Arco (then Atlantic) decided that my oil furnace system, Jet-Heet, was the way to fight against the use of gas heat in new housing developments because Jet-Heet's small, flexible, insulated ducts were both efficient and easy to install. In this case we were already selling a good volume but, enticed by the prospect of working with those companies, we dropped our prices, raising our breakeven level above our existing sales. We also had to provide air conditioning, water heating, registers of many types, and different sizes of everything—a big investment.

Neither Exxon nor Arco sold the system to more than one housing development. We never reached our new breakeven level, and severe losses almost sank us. We had to sell our line to another firm, which has done very well with it.

Don't try to move too fast. If you go the OEM route, be sure to diversify your marketing methods.

EXCLUSIVE NATIONAL DISTRIBUTORS

Another way you can go in marketing your invention is to contract with a marketing firm to be your exclusive distributor. This means that the product will be sold under the distributor's name and control and you will, in effect, be the manufacturer and engineer, perhaps shipping the product to the distributor's customers as well.

Classified ads are useful in finding a marketer. Many trade journals accept them and good marketers are on the lookout for them. They can also be used effectively in Sunday newspapers. Simply put it like this: "Exclusive national distributor wanted for new patented product in _____ field."

Such arrangements are usually nationwide, so that the distributor can enter the national trade shows, do public relations and advertising in national trade magazines, and arrange for chain outlet sales, service arrangements, and sales representation by people handling other national lines. Attempts to restrict the distributor will not work for either of you. The distributor is dependent upon you for many things and is not likely to undercut you.

You have three major protections against the distributor's

buying elsewhere in an arrangement of this kind: your patent, your contract, and the drop-shipping you do for them. You may not have the strongest patent in the field, and it may not keep competitors from selling a similar product, but it is very strong against someone you have under contract who has agreed to rely on it and with whom you have been doing continuous business.

Drop-Shipping

Drop-shipping is an arrangement whereby the product is shipped against individual orders to many customers from a single point separate from the advertised office or plant. It is a practice that has advantages for both parties to a national distributor agreement.

The distributor does not have to carry inventory, provide for trucking deliveries or shipments, handle merchandise, or take the associated insurance and security risks. This leaves him free to take only sales office space, with no warehousing, and simply say that all shipping is done from his plant.

As the manufacturer, you already have an operation well suited to warehousing and shipping. You can become much more informed about the practical aspects of the business. In addition, the list you have of all the customers is a strong protection for you; if the distributor makes arrangements for another source of supply, you can sell to his customers directly and at lower cost.

You will, of course, be taking on added expense when you supply the drop-shipping service, and the price per unit that you set should reflect this expense. When you bill the distributor, bill "as warehoused," not "as shipped." The distributor will still save money over what it would cost him to warehouse and ship himself. Developments such as COD shipments, customer pickups, spare parts, service arrangements, warranties, and so on will have to be ironed out and should be anticipated in the contract.

The Exclusive National Distribution Contract

The principal things to be covered in the contract are territory, quantities, price, warranty, shipping arrangements, the distributor's right to buy elsewhere (page 152), performance, terms of payment, design control, quality control, cost of tooling, patent rights, default, termination, infringement, defense of suits, and arbitration of disputes. It is an involved agreement and requires a lot of thought. As in most such agreements, the performance depends largely on the good faith of the people involved. It is important that the parties be compatible and not too far apart geographically, so that frequent meetings can occur. Each party has a real interest in seeing the arrangement work out, so the key element is to build in an opportunity to iron out problems, preferably as soon as they occur.

For example, if a product has been shipped to customers for months even though, because of some availability problem, it does not incorporate the agreed design details and if the distributor is not fully aware of this, there may be trouble. The customers complain, sales are lost, and the manufacturer won't take responsibility. A phone call and a follow-up note are needed long before the problem gets to this stage. Sales organizations need time to prepare customers for changes. Minor things easily become major in this sort of close yet separated relationship.

My company had such a relationship with a distributor for twenty-five years, one that worked out well although it was not always peaceful. We made the Roll-A-Grill hot dog cooker and the Roll-A-Grill Corporation of America marketed it nationally, and sometimes even exported it. All the issues mentioned above were covered in our twelve-page single-spaced contract and many amendments were added to it over the years. This was done without outside legal help because it was in essence a contract between two individuals, the two presidents, who had to be comfortable with it and have it express their attitudes.

When, in a contract, you agree to agree in the future on some point, a lawyer will say it means nothing; the parties involved, however, may well be guided to a solution by the contract's wording. If you hire a lawyer to draw up a contract, you'll be legally protected against any eventuality regardless of whether the contract reflects the pesonalities of those involved. People change, they say; they come and go. Entrepreneurs, however, are different in that their personalities are the dominant factor; if they are no longer going to be involved in the deal, the business agreement is not likely to survive anyway. If you hire a lawyer to prepare your distribution agreement, don't fuss if it isn't written just as you'd like. You can always write it yourself and let the lawyer point out the risks you've created. Then what you do about them is a business decision.

The Right to Buy Elsewhere

Here is a hint that may be helpful in relieving the distributor's concern over what will happen if the manufacturer's price is arbitrarily set too high. Give the distributor the right to buy elsewhere if (1) he presents a bona fide quotation more than 5 percent below the manufacturer's price, (2) he gives the manufacturer thirty days to revise his price to within 5 percent of the quoted figure and it isn't done, and (3) he pays an agreed royalty. This has worked well for me and saves endless complications over the cost of each item of material, labor, and overhead. Maybe the 5 percent should be 7, 10, or 2. That's for negotiations to determine.

TRADEMARKS

The ownership of a trademark for your product is most important. This becomes the product's chief protection in the marketplace and its chief factor of recognition. Since a distributor will be using his company name you should arrange to own the product trademark and license it out. This will give you another protection in your business arrangement or, indeed, in any marketing plan.

Picking a trademark, like picking the title of a book, is difficult. It involves time, long lists of possibilities, many people's suggestions, good taste, and analysis of what you need to convey. You don't want it to be too descriptive or misdescriptive, as it will then be weakened and may not be registerable. It should always be used as an adjective rather than a noun, so that it won't easily become a generic term, the common name for the product, as happened with the trademarks "aspirin" and "escalator."

"Kodak camera" is perfect because the word "Kodak" is neither descriptive nor misdescriptive; it is unlike any other word and modifies the generic word "camera." "Freon" is not descriptive, but it is rarely used in the combination "Freon refrigerant" and thus tends to become generic. The term "Thermos"—as in "Thermos bottle," was descriptive and even though it was intended for use as an adjective, it was commonly used as a noun. Thus it was weakened, and was eventually lost as a trademark.

Check the *Thomas Register* and trade directories for lists of trade names, which will help get you thinking. Start your friends and business associates making lists. My company's latest trademark, LEVLOAD ice banks, was gleaned from 200 possibilities; it is not descriptive of the cooling storage product yet suggests a connection to its off-peak use. It will always be used with the generic term "ice banks," which will describe the product and keep the trademark a trademark.

Try to make your trademark recognizable and memorable. You might, for example, make it an acronym. I suspect "Freon" came from "fluoride refrigerant," with the "on" added to make it sound scientific (as in "electron"). The misspellings commonly used, such as "LEVLOAD," are an attempt to make the word recognizable as a trademark and memorable because of the error.

There is a great deal to be learned about choosing, registering, and protecting a trademark. In Appendix H ("Recommended Reading") at the back of this book, you will find several publications to consult for further information.

THE IMPORTANCE OF MARKETING

It makes inventors unhappy to hear that marketing is the most important element in a new product's success. I console myself by saying it only seems that way because it's the last link in a long chain and therefore the most noticeable one. All the other links have had to be successfully forged to get to the last one. It's a pointless argument. But when you're in the last phase, marketing must be the dominant factor. A weak final link can break any chain.

The Roll-A-Grill hot dog cooker was exhibited at twenty-six trade shows a year, one every two weeks, to get it going. That's the kind of dedication the marketing and sales personnel must have. If they don't, make a change. Don't let someone add your invention to their line and not push it. There are always people who will do a good job for a good product if you try hard enough to find them.

CHAPTER 10

MARKETING IT YOURSELF

As everyone is aware, goods are sold in many different ways: retail outlets, coupon sales in a magazine, the salesman at the house or office, the business-to-business phone call, the service call, and direct mail are only a few of them. If you have decided to market your product or service yourself, you must go a step behind the final sale and examine the various ways goods get to the market and where your product fits in.

Because it is an invention, your product may or may not have a clear way to go. There are established marketing channels for every type of product, but sometimes—due to an invention's unique qualities, its degree of innovation, its cost, the personalities involved, or other factors—an unusual channel will work better.

The three basic modes of marketing are direct, wholesale, and through a manufacturer's representative. Direct marketing includes direct mail, house-to-house selling, catalog sales, and any other kind of solicitation or advertising where the direct-mail marketing house or the manufacturer is responsible for the customer contact. Wholesale involves a middleman who takes on functions such as buying and stocking of goods, distributing to retail outlets, financing the purchaser's credit, selling to businesses for resale or business use, selecting and training dealers, and so on. A manufacturer's representative generally works on commission and sells to clients that require special care and explanations (e.g., consulting engineers, architects, institutions, or government). There are all sorts of overlapping marketing methods, including those of stocking reps, who buy and stock the product themselves; wholesale buyers who sell at retail; sellers who take goods only on consignment; and distributors who give extensive special treatment.

DIRECT SALES

By "marketing it yourself" I do not mean that you must do it all. Quite the contrary; the help is there for the asking. The key question is what is the *best* way to sell your particular product. Don't think that if you have limited funds, you'll have to sell by direct mail. Actually, that can be the most expensive technique unless you have other products to sell at the same time. There are many direct-mail houses that might take on the assignment. They have proven mailing lists and they know what will sell. Others use newspaper or magazine ads. They handle the business in your name; you do the shipping and collections and pay the house for its service. Of course, you can also purchase mailing lists directly from many firms.

A lot of direct-mail selling is done when monthly credit card or other billings are sent to customers. Utilities sometimes send them, as do oil companies. Their large volume, however, results in low payments to vendors. Direct mail generally works on high markups, often three and even four times factory costs.

Direct mail is not the best marketing method for every

product. My company took a series of ads in *Esquire* magazine for our invention of a solar suntan box, a large fold-up, foil-lined, open-top box you can put out in the yard and lie down in in your bathing suit. You could beat the season, because even on a cold, sunny March day you'd feel perfectly warm from all the reflected sun. We sold several hundred of these boxes in early spring but none during the rest of the year. Then we were sued by a woman who felt humiliated when her box blew away in a gust of wind after she had unwisely doffed everything. Now, if we had been selling the box to dealers, they would have leveled out our sales by carrying some inventory and soothing the lady's pride in some way. (Incidentally, we gave her her money back and she dropped the litigation. I always wondered if she would have appeared in court and how she would have been dressed.)

A perfect direct-mail item is the Blowpoke, a fireplace tool that combines a poker with a tube through which you can blow to stimulate the fire. Stu Burt, its entrepreneurial owner and manufacturer, sells it through direct-mail catalogs primarily in New England, where fireplaces and wood stoves are common. Now at the age when he'd have to retire from a big company, he has a steady part-time business and is the envy of his peers.

PRICING

More mistakes are made in pricing than any other part of marketing. There are many temptations: raise prices because you've got something no one else does; lower prices and make it up with higher volume; meet your competitor's price so he won't get an advantage. These are games to be played when the product is a mature one and you know exactly what the costs are.

As an entrepreneur putting out your untried product, you

will be wise to stick to more conventional procedures and to set price on the basis of cost. A rule of thumb that many a well-managed company uses is to price on the basis of double the sum of the direct material and labor cost. This is simple to calculate and keep track of. If material cost goes up $1, the price goes up $2. If you give a labor increase of 50 cents an hour, increase the price by $1 for every direct labor hour. This avoids the "after you, Alphonse" routine, during which you and your competitors lose money waiting for someone to chicken out.

This doesn't mean you should change prices every month or even every quarter. But when you do, let your customers know in advance when the price change will take effect. Surprise is a poor idea in marketing.

You should also take into consideration any wholesale and retail markups that might have to be made. These are discussed in the following section.

WHOLESALE/RETAIL

Textbooks, college courses, and business schools by the thousands teach the workings of the American wholesale distribution technique. Each category of product calls for different ways of setting up the wholesale/retail relationship, so it is impossible to offer specific knowledge or advice related to your particular interest. However, there are some generalities that are pertinent to inventors:

1. When the product is new and different, the groups who will be interested in selling it are likely to be new and different themselves. If your product is something innovative in an established field, its problem will be credibility, and you will need an established organization behind it. If your product

is a rock music instrument, a new sales group is good because you need popularity, not credibility. But if you're selling a new solar water heater, you need the most reputable dealer in town.
2. Generally a wholesaler marks up the product price by 20 percent or so, sometimes less, sometimes more, depending on whether his is primarily a warehouse operation with just a few functions and employees or a distributorship with many responsibilities. The dealer or retailer usually marks the product up again by 30 to 50 percent. This makes it tempting for wholesalers to portray themselves as distributors and promise to get you dealers all over the state, while actually all they do is become "long discount" dealers themselves, buying as distributors and selling at retail. Unless your distributor is an experienced wholesaler, this is bound to happen. It may not be all bad because he'll be highly motivated and may move a lot of product, but you should restrict his territory sharply and expect the same thing to happen elsewhere.
3. Leave room in your pricing for both wholesale and retail markups, even though you might not plan to use them both now. Be assured that you will need to build both of them in before long, and it's hard to raise established prices. Use a multiplier on your list price instead of quoting a dollar amount. This gives you the flexibility to meet any situation. For example, if your product's factory cost is $1, make your list price $4, your dealer multiplier 0.60, your distributor multiplier 0.48, and your OEM multiplier 0.42, which will bring you $1.68.
4. Remember that your wholesaler's most important function is to carry the customer credit. This is invaluable to you, although it may not seem so at

first. The functions you may think of as primary, such as dealer selection and training, are rarely initiated by the wholesaler. Dealers are found through advertising and public relations, and their training will be primarily factory-sponsored.

5. A wholesaler won't carry your product if it doesn't make money for him, so it must sell well. But he won't work at selling it if he can do better with his other lines. It's a sort of Catch-22 that can be resolved only by your doing the initial selling and bringing it through him. It works particularly well if you can sell to his customers. Try and get the names of some of his customers and go see them. This kind of leg work is part of building a business.

6. Even if your product is not that great and the markups you allow are not too generous, you can win a good customer if you can be superresponsive to his letters, orders, or calls. One-day service or even any sort of immediate response is rare and greatly appreciated. Get to know the person to call and quickly get onto a first-name basis. You don't have to be a salesman to do this, just a responsive supplier who uses the phone.

7. Cooperative advertising and sales literature are needed and will be requested by any distributor who is going to do a good job for you. There are many ways to handle it and you should find out what is usual in your industry. It's expensive but necessary. Try to get the cost to you related to sales as a credit against some percent of purchases. You might have to pay for the first ad instead of crediting it to show that it can produce leads. Bill the distributor for sales literature, after giving him an initial free supply. Remember the adage, "It's only worth what you pay for it." Designing appropriate

literature that is not expensive is always a creative challenge for the small businessman.
8. If the wholesaler sells over the counter or by using traveling salesmen, design and supply an intriguing model, handout, or poster that will give the customer a quick visual understanding of the product and make it memorable. For example, the Klixon snap-disk thermal motor protectors became popular after people saw the jumping disks, which snapped as they heated or cooled.

REPS

Manufacturers' representatives are professional independent salesmen who handle a number of related lines in a given area and sell largely to contractors through consulting engineers. They usually sell products that require more training, explanation, application, and expertise than wholesale products. Often these goods are known as specialty items as opposed to commodities.

Reps like to carry new items that appeal to the same customers they are already calling on for their other lines. They particularly like items that require buying another one of their products. Reps call primarily on consulting engineers, design builders, manufacturers, and large contractors, but they often work with wholesale distributors as well. In cities like New York, where so much building is done, some reps call only on a few large consulting engineering firms, while others call on contractors, industry, and smaller engineering firms.

Commissions depend on how hard it is to make a sale. Established reps who primarily take orders from firms they've known a long time may get as little as 6 or 7 percent. However, a new product in a new market may well pay 15 percent or more. The rep has to make money, and he won't handle a product unless he can.

Contracts with reps usually cover matters such as products, territory, sales plan, compensation, technical service, expenses, relationship, trademarks, prices and terms, the sale agreement that is made with the customers, changing models, and terminations. It has generally been found that rep agreements work only when both parties are happy with the relationship, so the terms of the contract tend to be relatively loose and subject to either party's right to terminate the agreement in thirty days.

Manufacturers' reps are independent businessmen who like what they do. They will give you quick and frank appraisals of your product and its market, telling you what must be done to make it succeed. They go to trade shows regularly and have a good feel for the national market and coming changes. You can often locate reps through trade associations. If you have an invention that will boost someone else's product, you may be able to get rep lists from that person.

One inventor who made good use of reps was Dr. William McIntire, a brilliant solar scientist at Argonne National Laboratory near Chicago. He had heard that because of a natural gas shortage he would not be able to buy a decorative outdoor gas lamp for his front gate. Therefore he invented and built a flickering electric lamp that looked like the gas light he'd wanted. Encouraged by friends, he built more; finally, he and his wife set up an assembly and packaging line in their basement.

To sell the lamp, they tried advertisements with coupons, but orders came slowly. Discouraged, McIntire tried to sell his invention to a lighting company, but they felt it was too different from their line and were not interested. On an inspiration, he asked, "Would you mind if I contacted the reps who handle your product?" "Not at all," they replied.

"Here is a full list." The reps were delighted to handle it and sales boomed until gas became plentiful again.

The principal sales activity of a rep is building for the future. This may mean getting the product written into specifications by engineers which will later be bid on by contractors—the "bid and spec" market. Or it may mean getting a manufacturer to design your component into a new product that will be coming out later. Or it may mean getting a wholesaler to stock it when his present model is exhausted. All are made "on the come" with present expense but not present income. It is important to realize that a rep is making a big commitment when he takes on your product. Manufacturers should honor this.

Obviously, markets differ greatly geographically. Urban vs. rural, big city vs. small, long shipping distances vs. short, economically strong areas vs. economically weak—all these considerations affect how a rep must operate to serve the market and make money. Some reps become warehousers of some items so they can supply goods quickly; they are called stocking reps. Some buy and resell as a distributor does but still focus on just one or more product lines. Some are really full-line distributors but treat only one line as a priority item, for which a salesman makes calls. The principal thing you want to look for is that your representative, whatever he calls himself, will make calls on prospective customers.

You will undoubtedly find when marketing an innovative product that you have to try out a great many ideas on a great many different people before finding the ones that work for you. Some contacts will like what you've got, but, as with any untried invention, many more will want to wait and see. Follow up personally on any promising leads, and don't be discouraged if it takes a while for your product to build its sales.

CHAPTER 11

NEW FIELDS TO CONQUER

An invention starts with and is based upon its creator, or perhaps several creators; from the time it is put down on paper, all other operations that may contribute to its eventual success are services of one form or another. I can hear the howls now, for everyone thinks his or her function in business is most important. The financiers think money is basic. The marketers think selling is basic. Manufacturers, of course, think the making of the product is where it all begins. They all have their different points of view. The unique element however, is the inventor, and because of this he should be more honored in his creative work, as are authors and artists; in being so honored he will have the courage to do more. For without the inventor and the business he stimulates, America will become a second-rate nation, full of capable service organizations of every kind ready to do what they are hired to do, a nation without purpose. Of course, facilities, money, sources of knowledge, and other services are very helpful—they are just not more important than the creative spirit.

If I listed here, as has often been done, specific technical solutions that are needed, such as how to decontaminate

New Fields to Conquer

ground sprayed with Agent Orange, I would be viewing the inventor as the provider of a service, as if he were an institute or a government laboratory. This is not the best function of the free-lance inventor.

The inventor must see the need in an undirected way and find that unique solution that defies coercion. If a good invention could be "managed" into existence, it would exist already.

In Japan, for example, the corporate structure puts great distance between managers and engineers. There, inventions are managed that are largely improvements, not true innovations. As Ray Brill of the House Science and Technology Committee observed at a June 1982 conference for small technology-based firms sponsored by the National Science Foundation, for every 10,000 people, Japan has 400 engineers to our 70, 3 accountants to our 40, and 1 lawyer to our 20. No wonder the Japanese make such industrial progress; but in so doing they lose the independence of the free-lance "new to the world" inventor. It is the free thinkers who create the new industries.

Look for fields that need help, that are technically obsolete, that are bound up in custom, that haven't had a major change in a long while, and that benefit mankind. It doesn't make much sense to try to invent a new automobile, because that field is already a focus of technical brainpower and constant change, and thirsty industrial giants are ready to pounce on any great new idea that comes along. The aircraft industry is similar.

But take, for example, the fire fighting and prevention industry. It has seen relatively little change for 100 years, few companies are involved in it (and those that are are not large), the market is tightly bound by tradition, and the technology is not advanced. Thousands of victims die every year, arson is rampant, and business and building costs are vitally affected. Here is an industry ripe for innovation.

Suppose an inventor came up with a liquid kept in pressure tanks that, when released, would create tremendous amounts of a gas that would sustain life but not fire, such as a gas containing 7 percent oxygen. This could be released in seconds through pipes in all the rooms, putting out a fire with no damage; it would ventilate out the smoke and would be much cheaper to install than a sprinkler system, since it would require just one open pipe for each room with no connection to the water supply.

No one in the fire prevention business is likely to work on an idea like this, since it would make their existing products and services obsolete. Codes, standards, enforcement personnel, and fire and building authorities would deter anyone without the boundless enthusiasm and ego of a free-lance inventor.

So find a place where an invention is needed—where the industry is technically neglected and the potential reward big. Go through the Yellow Pages and industrial directories; attend trade fairs. Stick to what you know about technically or get a coinventor to provide the knowledge you don't have. Coinvention is probably the best potential source of meaningful innovation for the coming years.

My son Mike is a scientist at Lawrence Livermore Laboratory in California and has worked under Dr. Edward Teller, famous for the hydrogen bomb. The most remarkable thing about Teller, Mike says, is how he can keep over 100 projects going at once. Each project is developed by a separate little group or individual, and Teller visits them all in turn, stimulating them with what has come to be known in the lab as a "Tellerism": an idea, usually incomplete and often totally impractical, that has enough original thought behind it to stimulate the project scientist to go in new directions.

Teller once told Mike that he might help the smog problem by throwing tons of soot out of an airplane into the

New Fields to Conquer

smog. The black soot would absorb solar energy, be heated, and rise, carrying the smog with it. The idea helped Teller's people to generate the atmospheric computer models that tell what the atmosphere does.

You can generate your own Tellerisms. An inventor's mind can't help but tackle many problems, most of which are dropped in favor of better ideas. It's good to juggle many mental balls at once so you can go from one to the next without getting discouraged. Each time around, you understand the problems better; by dropping some and tossing others higher, you finally arrive at the one you want to concentrate on.

A FEW CASE HISTORIES

Sailboarding is the world's fastest growing sport: more than a million boards have been sold in Europe, and in the United States dealers boast over 50,000 sales per year. The business is growing by 50 to 75 percent annually.

Three inventors are responsible for this phenomenon: S. Newman Darby of Wilkes-Barre, Pennsylvania, and later, independently, coinventors Hoyle Schweitzer and Jim Drake in Torrance, California. Darby did not file for a patent but, in 1965, published an article instead; he also made custom sailboards. Schweitzer and Drake invented their own version in 1968. Schweitzer bought out Drake, founded a company—Windsurfing International, Inc.—and mass-produced, sold, and licensed their streamlined fore-and-aft rigged version all over the world.

The key idea was that the spar to which the sail was attached was only pivotally attached to the board, not rigidly as in sailboats or canoes. The spar stays up only because someone holds it up. Darby invented this but didn't have the money to patent it. If he had known how, he could have done it himself; since he didn't, that protection was lost forever. Schweitzer and Drake also invented this feature, not knowing of Darby's earlier work, and they also worked out a better board and rigging, which made the whole idea catch on.

The validity of Schweitzer and Drake's patent has been the subject of ten years of legal battles. Was it truly inventive over Darby? Many infringers have entered the field and Schweitzer is suing them, demanding they take licenses and pay him royalties. Schweitzer applied for a reissue patent in order to broaden his claims and make them more valid now that he had learned of Darby. That brought a difference of opinion between the Patent Office Board of Appeals and the Examiner. Partly, it's the tendency to favor the inventor who gets a patent and creates an industry and employment, and that's as it should be. The Constitution authorized patents and they were put in the Department of Commerce to promote commerce, not to be strict legal spoilers. An initial monopoly provides the incentive for entrepreneurs to build new industries, and in that way this country is truly unique. As supply-side theorist George Gilder writes in his book *Wealth and Poverty*, "Entrepreneurs are fighting America's only serious war against poverty. The potentialities of invention and enterprise are now greater than ever before in human history."

Although Schweitzer, who built an industry, is now losing money and has spent half a million dollars on legal bills, he made a great deal of money in the past, particularly in foreign licenses. The Patent Office has validated some of his new claims, but not the broadest one. Competitors are increasing. If Darby had patented the idea broadly in the beginning, he and Schweitzer together would have made a fortune. As it was, though, they both did well in their businesses, thousands of people gained employment, and sailboarding became a sport enjoyed by hundreds of thousands.

New Fields to Conquer

Another team of inventors who struck it big—very big—were Steven Jobs and Stephen Wozniak of Apple Computer, Inc., in Cupertino, California. Both men—old friends who had once played together with an illegal telephone device they'd made allowing free long-distance calls—were bright college dropouts who worked for Atari and Hewlett-Packard respectively. Wozniak, in his spare time, designed and built a small, easy-to-use computer; Jobs saw the potential of it, redesigned it in trim plastic, got the necessary professional help, and raised the money to get going.

In 1977, the Apple Computer had sales of $2.7 million; in 1980, $200 million. Fabulously rich, Wozniak went back to college, while Jobs is fighting Radio Shack, Xerox, and now IBM to stay on top in personal computers. In the last three years, hundreds of new microprocessor companies have been privately funded under the appealing label of "high technology." Although this kind of trendiness cannot last—there will soon be an oversupply—thousands and even millions of jobs have been created and many inventors have tasted success.

High technology is not limited to silicon chips and lasers, as the press would have us believe. It is at the forefront of technology in any industry or science, the new and innovative that no one else has done. It is accessible to the inventor if he applies himself to work coming from the scientists. The combination of science, invention, and a promotion-minded partner are what leads to real money and success.

The hottest product of today and tomorrow is the computer. If I told you that Edison invented the computer, you wouldn't believe it. You would say that I've got my generations—if not centuries—mixed up.

Nevertheless, it's true. Theodore Edison, the youngest son of famous Thomas Alva, came up with some of the key elements of modern-day computers through work he was doing for the Signal Corps during World War II. For example, the shift register that takes the input and shifts it into memory is completely covered by one of his twenty-nine patent claims.

At the beginning of the war, it was widely feared that the Germans would mount transatlantic air raids on America's East Coast. Therefore the Signal Corps asked Edison to develop a system that would plot the location of incoming planes visually on a display map. It was eleven years after his father's death, and Theodore Edison then owned and ran a small development company, called Calibron Products, on the top floor of one of the Edison Company buildings in West Orange, New Jersey. It was here that some winners of the famous Edison scholarship were given jobs after being put through college. Calibron turned out many fascinating exhibits for the Museum of Science and Industry in New York and for the 1933 World's Fair in Chicago.

The invention that had caught the Signal Corps' attention was a stereoscopic map-plotting instrument developed for the American Geographic Society to counter a German device. From aerial photographs taken every half mile at a height of 15,000 feet, it could provide a contour land map accurate to within 6 feet. This was in 1940, twenty years before the U-2's.

In 1941 there was no radar, and planes had to find their way and detect other aircraft by using directional radio waves. The idea was to place large directional antennas up and down the East Coast to catch signals from planes over the Atlantic. These signals were to come into a master device that would indicate the location of the planes on a large wall map in a control room.

Edison worked seven long days a week, almost ruining his health. In 1944 he was nearing completion of the system when the Signal Corps looked it over and told him, "That's fine, but we don't need it any more. We have radar." After the officer explained what radar was, Edison asked, "Why didn't you tell me that you knew about this two years ago? It would have saved me unbelievable work and strain." Their answer: "It was a secret."

Understandably discouraged, Edison stopped inventing and developing products after that, but he went to work in characteristically meticulous fashion to patent all the things he had created in his mammoth job for the Signal Corps. The result was a "jumbo" patent, #2,724,183, consisting of sixty pages, thirty-one drawings, and twenty-nine claims. When I asked him whether he had really done everything on that patent application himself, he said, "No, I goofed. I didn't know you were supposed to put shading on the drawings. So the Patent Office had to do that."

And so the computer was born. Many other people later developed the same things and much more. Edison's patent was in the Patent Office for ten years. It survived seventy-three citations of prior art, all legally tested against his inventions. His patent was intensely complicated, pushing the limits of human perception and understanding. Although his father lost out on credit for inventing the telephone (a close call), let's hope Theodore Edison will receive due credit for his contribution to the computer and the electronic world of today and tomorrow.

Did he make any money out of it? Very little. He made a cold call on IBM one day when he happened to be in New York City. He asked to see the president, but settled for the legal V.P. When he saw the patent and the Edison name, the V.P. asked how much it would cost IBM. Edison replied, "$20,000 for a nonexclusive paid-up license." "I'll get a check," said the astounded V.P., and Edison walked out with the check in his pocket. Later he also sold a license to Western Electric. "I didn't even get $1 an hour," said Edison, "but maybe some good will come of it."

THE FAILURES AND THE SUCCESSES

An inventor has a batting average, just like a baseball player. Invention is a tougher game, however, so the top hitters, the Edisons, bat about .100, whereas in baseball a Ted Williams could reach .400. The inventor must be able to handle a multitude of strikeouts and walk away from the plate without being permanently discouraged.

How do you know when you've struck out, since there is no plate umpire? Most likely you will know through an analysis of a combination of things, such as the rejection or weakness of your patent, the lack of enthusiasm of established marketers, the margin of profit between cost and required selling price, and the technical performance risk involved.

You have to weigh whether or not you are experienced enough to find new patentable features, excite marketers, reduce cost, and improve performance and reliability. All this can and does happen. Negative answers defeat corporations more easily because they are more limited in the alternatives they can adopt. Inventors have more staying power, often because they have more latitude in problem solving. If the invention doesn't fit one market, for example, the inventor can try another. Perhaps you've tried to be too different in your invention, and one innovation would be more suitable than three. Get all the advice you can, but it's up to you to decide.

But it's important to distinguish between intuitive conviction and stubborness. There will be other inventions, maybe some that hit the same market, which may be easier for you to do. Profit from your mistakes. Walter J. Cairns, a vice-president of Arthur D. Little, Inc., reports that a successful license agreement yields the inventor $40,000 per year on the average; he concludes that an inventor is not likely to get rich. However, this ignores the fact that a patent lasts seventeen years. And the success of a new product is likely to grow over quite a period of time, yielding more royalties every year. Many successful patents reach their largest earning capacity only when they are about to expire.

Even more important, a successful license leads the inventor into many opportunities for other inventions and licenses which would never have been available otherwise. He gains a reputation as a moneymaker. Consulting contracts become available and new license negotiations on other patents, or other uses of the primary patent, become easier to arrange.

Royalties from patents can be treated as capital gains under certain circumstances, whereas book royalties cannot. Thus the maximum tax bracket would be 20 percent, a real incentive. If you can earn $100,000 in royalties, you can probably keep $97,000; a person earning that in salary might keep only $69,000.

Foreign countries are getting more and more interested in licensing U.S. technology. Income from foreign patents on the same invention can double or triple U.S. royalties if the market is there. Although you may not have had the money for foreign patents within one year of your U.S. filing date, you can file improvement patents overseas later on even though they are limited. Also, even without foreign patents, foreign companies will buy the know-how with a large up-front payment that represents royalties.

Joint ventures can be established in this country or overseas; in these, you contribute the patent rights and the other party puts in capital. As a joint owner, your return comes both in a share of profits and in the increasing value of the ownership of the venture. This arrangement is common in many countries on U.S. patents, with the U.S. inventor having little responsibility.

Once you are successful with one product, other opportunities will open up. The world is waiting for good ideas from those who know how to use them. I hope this book will help bring you success, for we will all benefit from future inventions. May you be able to say, as Lee Smith wrote in *Fortune*, the two most exciting words an inventor can utter: " 'Eureka' (I found it!) and 'Epolisa' (I sold it!)."

APPENDICES

APPENDIX A

SOME IMPORTANT PATENTS ISSUED IN THE UNITED STATES, PRIMARILY TO INDEPENDENT INVENTORS, IN THE LAST 100 YEARS

Year	Inventor	Subject and Title of Invention	Patent Number
1882	William F. Ford	"Stethoscope"	257,487
1883	James Ritty; John Birch	"Cash Register and Indicator"	271,363
1884	George Eastman	Transparent photographic paper-strip film; "Photographic Film"	306,594
1887	Emile Berliner	Record, disk; "Gramophone"	372,786
1888	Oliver B. Shallenberger	Electric meter; "Meter for Alternating Electric Currents"	388,003

Source: "The Patent System," U.S. Government Printing Office

Year	Inventor	Subject and Title of Invention	Patent Number
1888	George Eastman	Roll-film camera (Kodak®); "Camera"	388,850
	John J. Loud	Ballpoint pen; "Pen"	392,046
1890	Ottmar Mergenthaler	Linotype®; "Machine for Producing Linotypes, Type-Matrices, etc."	436,532
1891	Almon B. Strowger	"Automatic Telephone-Exchange"	447,918
1893	Frederick E. Ives	Half-tone printing; "Photogravure-Printing Plate"	495,341
	Charles J. Van Depoele	Electric trolley car; "Travelling Contact for Electric Railways"	495,443
	Whitcomb L. Judson	Zipper®; "Clasp Locker or Unlocker for Shoes"	504,038
1895	Charles E. Duryea	Gasoline automobile; "Road-Vehicle"	540,648
	George B. Selden	Road carriage; "Road-Engine"	549,160
1896	Clement A. Hardy	"Rotary-Disc Plow"	556,972
	Tolbert Lanston	Monotype®; "Machine for Making Justified Lines of Type"	557,994
	Edward Goodrich Acheson	Production of Carborundum®; "Electrical Furnace"	560,291
1897	Simon Lake	Even-keel submarine; "Submarine Vessel"	581,213
1898	Henry Ford	"Carburetor"	610,040
1899	Ferdinand Graf Zeppelin	"Navigable Balloon"	621,195
	John S. Thurman	Motor-driven vacuum cleaner; "Pneumatic Carpet-Renovator"	634,042
	Charles G. Curtis	Gas turbine; "Apparatus for Generating Mechanical Power"	635,919
1900	Felix Hoffmann	Aspirin; "Acetyl Salicylic Acid"	644,077
	Francis E. Stanley; Freelan O. Stanley	Stanley Steamer; "Motor-Vehicle"	657,711

Some Important Patents

Year	Inventor	Subject and Title of Invention	Patent Number
1900	Valdemar Poulsen	Magnetic tape recorder; "Method of Recording and Reproducing Sounds or Signals"	661,619
1901	Henry Ford	"Motor Carriage"	686,046
1903	Clyde J. Coleman	Starting motor for auto; "Means for Operating Motor-Vehicles"	745,157
1904	Frank A. Seiberling; William C. Stevens	Automobile tire-making machine; "Machine for Making Outer Casings for Double-Tube Tires"	762,561
	Michael J. Owens	Glass-shaping bottle machine; "Glass Shaping Machine"	766,768
	King C. Gillette	Safety razor; "Razor" (title for both patents)	775,134 and 775,135
1905	John Ambrose Fleming	Two-element vacuum tube; "Instrument for Converting Alternating Electric Currents into Continuous Currents"	803,684
1906	Lee De Forest	Radio tube detector; "Oscillation-Responsive Device"	836,070
1907	Lee De Forest	Radio amplifier tube; "Device for Amplifying Feeble Electrical Currents"	841,387
1908	Lee De Forest	Three-element vacuum tube (Audion); "Space-Telegraphy"	879,532
1909	Leo H. Baekeland	Bakelite®; "Condensation Product and Method of Making Same"	942,809
1910	James C. Dow	Electrical insulators; "Insulating Material"	952,513

Appendix A

Year	Inventor	Subject and Title of Invention	Patent Number
1910	Charles Y. Knight	Sleeve-valve engine; "Internal-Combustion Engine"	968,166
	Fritz Habor; Robert Le Rossignol	Synthetic ammonia; "Production of Ammonia"	971,501
1912	Peter Cooper Hewitt	Mercury vapor lamp; "Apparatus for the Electrical Production of Light"	1,030,178
1913	William M. Burton	Cracking process; "Manufacture of Gasoline"	1,049,667
1914	Robert H. Goddard	Rocket engine; "Rocket Apparatus"	1,103,503
	Edwin H. Armstrong	"Wireless Receiving System"	1,113,149
	Mary Phelps Jacob (Caresse Crosby)	"Brassiere"	1,115,674
1915	Glenn H. Curtiss	Hydro airplane; "Heavier-Than-Air Flying Machine"	1,156,215
1916	Ernst F. W. Alexanderson	Selective radio-tuning system; "Selective Tuning System"	1,173,079
	Irving Langmuir	Incandescent gas lamp; "Incandescent Electric Lamp"	1,180,159
1917	Carl R. Englund	"Radiotelephony"	1,245,446
1918	Elmer A. Sperry	Gyrocompass; "Gyroscopic Compass"	1,279,471
1921	Harry Houdini	"Diver's Suit" (permitting escape)	1,370,316
1925	Irving Langmuir	High-vacuum radio tube; "Electrical Discharge Apparatus and Process of Preparing and Using the Same"	1,558,436
1926	Thomas Midgley, Jr.	Lead ethyl gasoline; "Method and Means for Using Motor Fuels"	1,573,846
1928	Jacob Schick	Electric razor; "Shaving Implement"	1,721,530

Some Important Patents

Year	Inventor	Subject and Title of Invention	Patent Number
1930	Clarence Birdseye	Packaged frozen foods; "Method of Preparing Food Products"	1,773,079
	Philo T. Farnsworth	"Television System"; "Television Receiving System"	1,773,980 and 1,773,981
	Albert Einstein; Leo Szilard	Refrigeration apparatus; "Refrigeration"	1,781,541
1931	George Lewis McCarthy	Microfilming camera; "Photographing Apparatus"	1,806,763
	Thomas Midgley, Jr.; Albert R. Henne; Robert R. McNary	Refrigerants, low-boiling fluorine compound (Freon®); "Heat Transfer"	1,833,847
1933	Edwin H. Land; Joseph S. Friedman	"Polarizing Refracting Bodies"	1,918,848
	Edwin H. Armstrong	Two-path FM radio; "Radio Signalling System"	1,941,066
1934	Ernest O. Lawrence	Cyclotron; "Method and Apparatus for the Acceleration of Ions"	1,948,384
1935	Charles B. Darrow	Monopoly®; "Board Game Apparatus"	2,026,082
1937	Wallace H. Carothers	Nylon; "Linear Condensation Polymers"; "Fiber and Method of Producing It"	2,071,250 and 2,071,251
1938	Vladimir K. Zworykin	Cathode-ray tube, television; "Cathode-Ray Tube"	2,139,296
1940	Chester F. Carlson	Xerography; "Electron Photography"	2,221,776
1943	Paul Muller	D.D.T. (dichlorodiphenyltrichloroethane); "Devitalizing Composition of Matter"	2,329,074
1944	Marvin Camras	Eight patents on Magnetic Recording	2,351,004–2,351,011
1946	James M. Sprague	Sulfonamide; "Diazine Compounds"	2,407,966

Year	Inventor	Subject and Title of Invention	Patent Number
1949	Otto K. Behrens; Joseph W. Corse; Ruben G. Jones; Quentin F. Soper	Antibiotic; "Process and Culture Media for Producing New Penicillins"	2,479,295
	Keith Dwight Millis; Albert Paul Gagnebin; Norman Baden Pilling	Ductile cast iron; "Cast Ferrous Alloy"	2,485,760
1951	James W. Clapp; Richard O. Roblin	Diuretic; "Heterocyclic Sulfonamides and Methods of Preparation Thereof"	2,554,816
1952	Herbert C. Murray; Durey H. Peterson	Corticosteroid; "Oxygenation of Steroids by Mucorales Fungi"	2,602,769
1957	Joseph Seifter; Anthony L. Monaco; Franklin Judson Hoover	Tranquilizer; "Anti-Excitatory Compositions"	2,799,619
1958	Robert C. Baumann	"Satellite Structure"	2,835,548
1961	Robert N. Noyce	Integrated circuit; "Semiconductor Device-and-Lead Structure"	2,981,877
1964	Richard Buckminster Fuller	Geodesic dome; "Suspension Building"	3,139,957
1967	Harry E. Thomason	"Apparatus for Cooling and Solar Heating a House"	3,295,591
1976	Sidney Jacoby	"Combination Smoke and Heat Detector Alarm"	3,938,115

DESIGN PATENTS

Year	Inventor	Subject and Title	Patent Number
1842	George Bruce	Typeface; "Printing Type"	Des. 1
1879	Auguste Bartholdi	Statue of Liberty; "Design for a Statue"	Des. 11,023
1904	George L. Gillespie	Congressional Medal of Honor; "Design for a Badge"	Des. 37,236
1948	William G. Bley	Lady's stocking; "Design for a Lady's Stocking"	Des. 151,732

APPENDIX B

PATENT SEARCHERS

The best files and references for patent searchers are kept at the U.S. Patent Office, in the Search Room and Scientific Library; therefore, this list includes primarily searchers in the Washington, D.C. area. Most attorneys have searchers in the Washington area do their preliminary search work.

National Inventors Foundation, Inc.
345 W. Cypress
Glendale, CA 91204

Inventors Workshop International
121 N. Fir St.
Ventura, CA 93001

Thomas Collins
SDC Search Service
System Development Corporation
2500 Colorado Ave.
Santa Monica, CA 90406

Invention Marketing, Inc.
1701 K St., NW
Washington, DC 20006

National Patent Trademark Co.
2208 Wisconsin Ave., NW
Washington, DC 20007

Stan Stanton's Search Service
927 15th St., NW
Washington, DC 20005

Vincent Lafranchi
8504 Oglethorpe St.
New Carrollton, MD 20784

Government Liaison Service, Inc.
1011 Arlington Blvd.
Arlington, VA 22209

Invention, Inc.
2001 Jefferson Davis Hwy.
Arlington, VA 22202

National Patent Search Associates
703 S. 23rd St.
Arlington, VA 22202

Washington Patent Office Searches
1011 Arlington Blvd.
Arlington, VA 22209

Woolcott & Co.
1911 Jefferson Davis Hwy.
Arlington, VA 22202

APPENDIX C

GROUPS THAT HELP EVALUATE IDEAS

Evaluations from these groups are either free or available at nominal cost.

California State University, Fresno
Bureau of Business Research & Service
Fresno, CA 93740

Stanford University
Innovation Center
Stanford, CA 94305

George Washington University
Innovation Information Center
2130 H Street, N.W.
Washington, DC 20052

Dr. Dvorkovitz & Associates
P.O. Box 1748
Ormond Beach, FL 32074

University of Illinois
Bureau of Economic and Business Research
408 David Kinley Hall
Urbana, IL 61801

The University of Kansas Center for Research, Inc.
2291 Irving Hill Rd.
Campus West
Lawrence, KS 66045

Arthur D. Little, Inc.
Acorn Park
Cambridge, MA 02140

Jackson State University
Bureau of Business and Economic Research
1400 J. R. Lynch St.
Jackson, MS 39217

Center for Innovation
Montana Energy & MHD Research & Development Institute
225 South Idaho St.
Butte, MT 59701

Groups that Help Evaluate Ideas

Franklin Pierce Innovation Clinic
Franklin Pierce Law Center
Concord, NH 03301

Product Resources International, Inc.
90 Park Ave.
New York, NY 10016

Refac Technology Development Corp.
122 East 42nd St.
New York, NY 10017

Battelle Development Corporation
505 King Avenue
Columbus, OH 43201

Carnegie-Mellon University
Center for Entrepreneurial Development
4516 Henry St.
Pittsburgh, PA 15213

Center for Private Enterprise & Entrepreneurship
Hankamer School of Business
Baylor University
Waco, TX 76073

University of Utah
Utah Innovation Center
Salt Lake City, UT 84112

Inventor Consultation Service
American Patent Law Association, Inc.
2001 Jefferson Davis Hwy., Suite 203
Arlington, VA 22202

National Bureau of Standards
Office of Energy-Related
    Inventions
Washington, DC 20234

Wisconsin Innovation Service Center
University of Wisconsin
Whitewater, WI 53190

APPENDIX D

PRODUCT DEVELOPMENT COMPANIES

Although the Yellow Pages list hundreds of product development companies, we have listed here some of those that have many years of experience in the business. Talking to them may help you to judge the quality of others.

Scale Models Unlimited
111 Independence Dr.
Menlo Park, CA 94025
415-324-2515
Models, display, architectural, technical; since 1961. Contact: Donald Nusbaum.

Model Builders, Inc.
6155 S. Oak Park Ave.
Chicago, IL 60638
312-586-6500
Industrial models, R&D, inventions; since 1950. Contact: William Chaffee.

Ceconn
Star Route #4, Box 2041
Branson, MO 65616
417-334-8307
Design, development, models, fabrication; since 1969. Contact: Charles Cooper.

Precision Forms, Inc.
Route 23 S., P.O. Box 788
Butler, NJ 07405
201-838-3800
R&D and design, machine shop; since 1956. Contact: William Sulski.

Production Previews, Inc.
29 E. 21st St.
New York, NY 10010
212-982-2290
All except electronic; since 1964. Contact: M. Chermack.

Design & Development Models
72 Pine St.
Oxford, PA 19363
215-932-9666
Appliances, toys, small mechanisms; since 1956. Contact: Riley Lohr.

HMS Associates Co.
2425 Maryland Rd.
Willow Grove, PA 19090
215-659-1923
Styling, engineering, prototyping; since 1951. Contact: J. Campbell.

Sail Engineering
P.O. Box 8439
Richmond, VA 23226
804-288-4819
Models, robots, electronics, engineering; since 1974. Contact: Vincent Serio, Jr.

Medalist Steel Products
2400 W. Cornell St.
Milwaukee, WI 53209
414-873-2010
Steel and aluminum fabricators, runs of 50 to 5,000; since 1916. Contact: Lou Burmeister.

APPENDIX E

SOME COMPANIES LOOKING FOR INVENTIONS

For addresses and complete information on the fields these companies are interested in, consult the *Thomas Register* or *U.S. Industrial Directory*. This list was compiled partially from information provided by the American Association of Small Research Companies.

Adolph Coors Co.
Aladdin Industries, Inc.
Allied Corp.
Allied Tube & Conduit Corp.
American Can Co.
American Cyanamid Co.
American Enka Co.
American Hospital Supply Co.
American Smelting & Refining Co.
J. T. Baker Chemical Co.
Balchem Corp.
Beckman Instruments
Becton, Dickinson & Co.
Black & Decker
BP Chemical
The Budd Co.
Chevron Chemical Co.

Ciba-Geigy Co.
Colgate-Palmolive Co.
Combustion Engineering, Inc.
De Soto, Inc.
Diamond Shamrock Corp.
Dow Chemical Co.
Eaton Corp.
Elf Acquitaine Inc.
Ethyl Corp.
Ex-Cell-O
FMC Corporation
General Mills
The Gillette Company
W. R. Grace & Company
Gulf & Western
Itek Corp.
Johnson, Matthey, Inc.

Some Companies Looking for Inventions

Kimberly-Clark Corporation
Koppers Company, Inc.
Arthur D. Little, Inc.
Merck & Company, Inc.
Monsanto
Nordson Corp.
Olin Corporations
Owens-Corning Fiberglas Corp.
Pfizer, Inc.
The Proctor & Gamble Company
The Quaker Oats Company
Raytheon Co.
Research Corp.
Reynolds Aluminum
A. H. Robins Company, Inc.
Rohm and Haas Company

Signode Corp.
Singer Company
SKF Industries, Inc.
SMC Corporation
Smith Kline Corporation
Sun Chemical Co.
Sunkist Growers, Inc.
Sybron Corp.
3M Company
TRW
Union Carbide Corp.
Uniroyal
Upjohn Company
Warner-Lambert
Westvaco Corporation
Xerox

APPENDIX F

SAMPLE LICENSE AND OPTION AGREEMENTS

License Agreement

AGREEMENT dated and effective as of _____, by and between _____ (hereinafter called Licensor), a corporation
_____(your company)_____
of the State of _____, having an office at
_____, and _____
                                          (the large company or whoever)
(hereinafter called Licensee), a corporation of the State of
_____, having an office at _____.

WITNESSETH:

WHEREAS, Licensor has been engaged in the development and manufacturing and marketing of systems and devices identified in Section 1.1 hereof as Licensed Products and has the right to grant licenses under the United States Letters Patent, foreign patents and patent applications listed on Schedule A.

WHEREAS, Licensee desires to obtain such manufacturing and marketing information and to make lawful use of the inventions of one or more of these patents and to that end desires to acquire the rights and licenses herein granted;

NOW, THEREFORE, in consideration of the premises, the licenses granted herein, and the agreements, covenants and conditions herein contained, it is agreed as follows:

ARTICLE I
DEFINITIONS

Section 1.1 *Licensed Products*—The term "Licensed Products" as used herein shall mean any _____

_____

manufactured by Licensee or its Associated Companies which
  (i) embody or the manufacture of which utilizes written information furnished by Licensor pursuant to this agreement; or
  (ii) embody or the manufacture of which utilizes any invention claimed in any of the unexpired patents listed on Schedule A attached hereto and any reissues, continuations or extensions thereof.

An example of Products is shown in U.S. Patent No. _____ entitled "_____". Licensed Products shall not include _____

Section 1.2 *Associated Company*—An Associated Company of a party to this agreement shall mean a company owned or controlled by the party or under common control with the party. A company is controlled by ownership of more than fifty percent (50%) of the stock entitled to vote for directors of the company or persons performing a function similar to that of directors. Licensor and Licensee are the parties to this agreement.

ARTICLE II
TECHNICAL ASSISTANCE

Section 2.1 *Documentation Supplied by Licensor*—As soon as reasonably possible, but in no event later than thirty (30) calendar days after the effective date of this agreement, Licensor shall furnish Licensee with all engineering, technical, manufacturing and test data and other information in reproducible form, including but not limited to prints, manufacturing drawings, process sheets, raw material and process specifications, manuals, drawings of manufacturing and test equipment, parts lists, test specifications, vendors' lists, and other writings in Licensor's and any Associated Company of Licensor's possession relating to Licensed Products, and piece parts, subassemblies, housings, components and elements relating thereto, which are necessary to permit Licensee to manufacture Licensed Products and such piece parts, subassemblies, housings, components and elements in Licensee's plants according to the best systems known to Licensor and to use such Licensed Products.

Section 2.2 *Marketing Information*—Licensor promptly after the execution of this agreement, but in no event later than thirty (30) days thereafter, shall furnish Licensee with full and complete information and data in its possession with respect to the marketing of Licensed Products including but not limited to one set of all descriptive, advertising and technical literature, published price lists, delivery requirements of customers, and information of like character relating to Licensed Products.

Section 2.3 *Instruction Provided by Licensor*—Promptly upon request of Licensee made at any time during the first three (3) months of this agreement, Licensor shall permit Licensee's technically skilled personnel designated by Licensee to make one or more visits to such facilities of Licensor as may be engaged in the manufacture of Licensed Products (with travel and living expenses paid by Licensee) for up to an aggregate of ten (10) man-days beginning with the effective date of this agreement, in order to inspect and be instructed in all engineering and manufacturing techniques and procedures according to the best systems of manufacture known to Licensor relating to the information and data to be supplied to Licensee pursuant to this Article II, and Licensor shall instruct them in and assist them in securing the complete know-how of fabricating parts, inspecting materials and components, testing parts and manufacturing and testing Licensed Products.

Section 2.4 *Licensee's Right to Use Information*—Licensee and its Associated Companies shall have the fully paid, exclusive, irrevocable right throughout the world to use all of the information and data furnished pursuant to Article II of this agreement.

ARTICLE III
GRANT

Section 3.1 *Grant of License*—During the term of this agreement and subject to the conditions hereof, Licensor grants to Licensee personal, indivisible, nontransferable, exclusive, world-wide licenses, with the right to sublicense its Associated Companies, under the patents listed on Schedule A to make, have made for its own account, use, lease, export and sell Licensed Products.

ARTICLE IV
COMPENSATION

Section 4.1 *Contract Payment and Royalty*—Licensee agrees to pay Licensor or its designee:

  a. Within ten (10) days after delivery of the information referred to in Sections 2.1 and 2.2, the sum of _____Dollars ($_____), which sum in no event shall be returnable;

b. During the term of this Agreement, a royalty of _____ Dollars ($_____) for each Licensed Product which is sold, leased or used by Licensee or its sublicensees until accrued royalties total _____ Dollars ($_____), at which time the royalty rate shall increase to _____ Dollars ($_____) per Licensed Product; provided, however, that when accrued royalties for any calendar year during the term of this Agreement total _____ Dollars ($_____), the royalty rate shall decrease to _____ Dollars ($_____) per Licensed Product on all additional Licensed Products sold, leased or used by Licensee or its sublicensees during the remainder of such calendar year.

No royalty shall accrue on Licensed Products solely because of an expired Licensed patent nor shall royalties be payable on sales of repair parts used to replace defective parts of similar type for the purpose of repairing a Licensed Product.

Section 4.2 *Minimum Royalty*—Licensee agrees to pay Licensor or its designee a minimum nonreturnable sum of _____ Dollars ($_____) per year during the term of this agreement, beginning with the effective date hereof, and Licensee agrees to pay the first such sum within thirty (30) days after delivery of the information referred to in Sections 2.1 and 2.2 hereof, and thereafter within thirty (30) days after each succeeding anniversary of the effective date of this agreement. Such sums shall be credited toward payment of royalties accruing during the term of this agreement pursuant to Section 4.1 of this Article IV.

Section 4.3 *Sales; Returns; Leases*—A Licensed Product shall be considered as sold or used and royalties shall accrue when such Licensed Product is billed out, or if not billed out, when delivered, shipped or mailed to Licensee's customer or to Licensee's internal operation or subsidiary which uses such Licensed Product. Licensed Products shall be considered as leased and royalties shall accrue when such Licensed Products are first leased out, delivered, shipped or mailed, whichever first occurs. Any Licensed Product in Licensee's possession as of the termination of this agreement on which royalty has not been paid by Licensee shall be considered for purposes of computation of royalty payment pursuant to this agreement to have been sold immediately prior to the termination of this agreement.

ARTICLE V
REPORTS AND PAYMENTS

Section 5.1 *Statement*—Within thirty (30) days after the end of each quarter year ending with the last day of March, June, September and

December of each calendar year during the term of this agreement, Licensee shall furnish Licensor or his designee with certified written statements showing Licensee's and its sublicensees' total number of Licensed Products sold, used or leased by them during the preceding quarter year, and the computation of royalties, and shall pay the royalties due pursuant to Article IV. Any credits pursuant to Section 4.2 for royalties previously paid shall be fully set forth on the statement and the amounts involved clearly shown. A similar statement shall be rendered and payment made within sixty (60) days after and as of the date of termination of this agreement covering the period from the end of that covered by the last preceding report to the date of termination. The first statement submitted under this agreement shall cover the period from the effective date of the agreement to the end of the quarter being reported.

Section 5.2 *Accounts*—Licensee shall keep for three years after the date of submission of each statement, true and accurate records, files and books of account containing all the data reasonably required for the full computation and verification of Licensee's and its sublicensees' number of Licensed Products sold and royalties to be paid as well as the other information to be given in the statements herein provided for. Licensee shall at its sole election either (i) permit Licensor or its duly authorized representatives or (ii) a Certified Public Accountant acceptable to Licensee and Licensor, upon reasonable notice, adequately to inspect the same at any time during usual business hours.

ARTICLE VI
TERM AND TERMINATION

Section 6.1 *Term*—This agreement shall continue in force until the expiration of the last-to-expire Licensed Patent, unless terminated by Licensee three years after the effective date of this agreement, or on any succeeding anniversary of such effective date, by ninety (90) days' prior written notice by Licensee to Licensor, unless sooner terminated by Licensor as hereinafter provided.

Section 6.2 *Termination on Default*—If Licensee shall at any time default in the payment of any monies due in accordance with this agreement or in fulfilling any of the other obligations or conditions hereof, and such default shall not be cured within sixty (60) days after notice from Licensor to Licensee specifying the nature of the default, Licensor shall then and thereafter have the right to terminate this agreement by giving written notice of termination to Licensee, and upon the giving of such notice of termination, this agreement shall terminate on the tenth day after such notice is given. Licensee shall have the right to cure any such default up to but not after the giving of such notice of termination.

Section 6.3 *Termination on Insolvency*—Upon the occurrence of any one of the following events:
  (i) The adjudication of Licensee to be bankrupt or insolvent;
  (ii) The filing by Licensee of a petition in bankruptcy or insolvency;
  (iii) The filing by Licensee of a petition or answer seeking reorganization or readjustment under any law relating to insolvency or bankruptcy;
  (iv) The appointment of a receiver with respect to all or substantially all of the property of Licensee;
  (v) Any assignment by Licensee of its assets for the benefit of creditors;
  (vi) The institution by Licensee of any proceedings for liquidation or the winding up of its business other than for purposes of reorganization, consolidation or merger;

this agreement may be terminated effective on the date of such event or at any time thereafter by notice from Licensor to Licensee electing to terminate this agreement.

Section 6.4 *Effect of Termination*—No termination of this agreement shall release Licensee from any of its obligations hereunder previously accrued or rescind or give rise to any right to rescind anything done or any payment made or other consideration given to Licensor hereunder prior to the time such termination becomes effective or to limit in any way any other remedy Licensor may have against Licensee.

Section 6.5 *Waiver of Default*—No failure or delay on the part of a Licensor in exercising its right of termination hereunder for any one or more defaults shall be construed to prejudice its right of termination for such or for any other or subsequent default.

ARTICLE VII
MISCELLANEOUS PROVISIONS

Section 7.1 *Patent Validity*—If a claim or claims of any patent licensed hereunder shall be held invalid by a court from whose decision no appeal is taken or no appeal or other proceeding for review can be taken, then such claim or claims shall, subsequent to the date of final unappealed or unappealable judgment, be treated as invalid and no royalties shall be due under this agreement for sales thereafter of Licensed Products covered solely by such claim. Licensee agrees to use its best efforts to protect the value of any patent licensed by this agreement, but it is agreed that there is no admission of validity by Licensee of any such patent.

Section 7.2 *Assignment*—This agreement shall be assignable by Licensee with disposition of the business to which licenses granted herein relate. In all other situations the agreement shall be assignable by Licensee with the prior written consent of Licensor, which consent shall not unreasonably be withheld.

Section 7.3 *Notices*—All notices, payments and statements given under this agreement shall be in writing and shall be deemed to have been properly addressed when, if given to Licensee, it shall be addressed to it as follows:

_____

_____

and when if given to Licensor it shall be addressed to it at the address given above, and in either case sent by registered or certified mail. The date of service shall be deemed to be the date on which such notice, payment or statement was posted. Either party may give written notice of a change of address, and after notice of such change has been received, any notice, payment or statement thereafter shall be given to such party as above provided at such changed address.

Section 7.4 *Construction*—This agreement is to be considered wholly executed and delivered within the State of _____, United States of America, and it is the intention of the parties that it shall be enforced, construed, interpreted and applied in accordance with the laws of _____ and Licensee hereby submits to the jurisdiction of the State courts of the State of _____ and the Federal courts situated within the State of _____ for such purposes. The headings of Articles and Sections in this agreement are included herein for convenience and shall not be considered in construing this agreement.

Section 7.5 *Extraneous Writings*—This agreement sets forth the entire agreement and understanding between the parties as to the subject matter of this agreement and merges all prior discussions and writings between them, and neither of the parties shall be bound by any conditions, definitions, warranties or representations with respect to the subject matter of this agreement, other than expressly provided in this agreement, or as duly set forth subsequent to the date hereof in writing and signed by an authorized person, or officer of the party to be bound thereby.

IN WITNESS WHEREOF, the parties have caused this agreement to be duly executed in their names by their proper officers thereunto duly authorized and their corporate seals to be hereunto affixed.

ATTEST: By _____
                                     Title:
_____

ATTEST: By _____
                                     Title:
_____

OPTION AGREEMENT

INTERNATIONAL PRODUCTS CO.
1000 City Highway
Houston, TX

XYZ Invention Co.   DATE: _____
219 Park Street
Smithton, OH
ATTN: Mr. John Smith, President

RE: Option on a License

Dear Mr. Smith:

The following shall constitute an agreement between you and us relative to our option to license certain rights to your product known as _____.

1. For the sum of Twenty Thousand ($20,000.) Dollars paid by us to you herewith, you hereby grant to us the irrevocable right and option for six (6) months from date hereof to license from you the exclusive rights to manufacture and sell the above product according to the terms of attached Exhibit A, "License Agreement."

2. You warrant to us that you have the right to grant licenses under the United States Letter Patent, foreign patents and patent applications listed in Schedule A.

3. You further warrant and represent that during the time this option is in existence, you shall not assign or encumber the patent rights in any way, nor make any commitment with respect thereto inconsistent with the terms of this option or the Exhibit A License Agreement.

4. Concurrently herewith we have both affixed our signatures to the Exhibit A License Agreement. Notwithstanding such signatures, the aforesaid agreement shall not become binding and effective except as provided in paragraphs 5 and 7 hereof.

5. This option may be exercisable by us at any time during the six-month Option Period by written notice sent by Registered Mail, Return Receipt Requested, on or before the expiration date, addressed to you at your address herein set forth.

6. (a) If we fail to exercise this option as aforesaid, all our rights hereunder shall cease on the date specified in Clause 1 above, and the agreement set forth in Exhibit A shall be null and void. (b) Within our sole discretion, however, we may decide to forego our option hereunder at a date during the Option Period. In such event, we shall notify you in writing, upon which event the terms of this paragraph 6 (a) shall immediately apply.

7. If we exercise this option as aforesaid, the rights and obligations specified in the agreement set forth in Exhibit A shall become immediately effective.

8. This agreement shall inure to the benefit of, and shall be binding upon your successors and assigns and our successors and assigns.

9. If for any reason, such as strikes, including strikes by applicable unions, boycotts, war, Acts of God, labor troubles, riots, delays of commercial carriers, restraints of public authority, or for any other reason, similar or dissimilar, beyond our control, we shall be unable to exercise our rights hereunder at any time during the Option Period or Extended Option hereof, then we shall have the right to extend the term hereof for an equivalent period, without any additional compensation to you.

If the above meets with your approval, kindly indicate your acceptance of the terms and conditions hereof by signing where indicated below and returning a signed copy of this letter to us.

                            Very truly yours,
                            INTERNATIONAL PRODUCTS CO.

                            By: _____
                                  President

ACCEPTED AND AGREED:
XYZ INVENTION CO.

By: _____
    President

APPENDIX G

THE SUPPORT NETWORK

INVENTORS' GROUPS

United Inventor Scientists
14431 Chase St.
Panorama City, CA 91402
Contact: Samuel Resnick

Inventors of California
250 Vernon St.
Oakland, CA 94610
Contact: Larry Brown

California Inventors Council
P.O. Box 2096
Sunnyvale, CA 94087
Contact: Barrett Johnson

Inventors Workshop International
121 N. Fir St.
Ventura, CA 91003
Contact: Melvin L. Fuller

or

36 W. 44th St.
New York, NY 10036
Contact: Tom Morrison

Inventor Associates of Georgia, Inc.
P.O. Box 95172
Atlanta, GA 30347
Contact: Joseph W. Buffington

Inventors Council of Hawaii
P.O. Box 27844
Honolulu, HI 96827
Contact: George Lee

Inventor's Association of New England
P.O. Box 3110
Cambridge, MA 02139
Contact: Dorothy Stevenson

Inventors Club of America, Inc.
121 Chestnut St.
Box 3799, Room 228
Springfield, MA 01101
Contact: Alexander T. Marinaccio

Minnesota Inventors Congress
P.O. Box 71
Redwood Falls, MN 56283
Contact: Raymond Walz

MidWest Inventors Society
P.O. Box 335
St. Cloud, MN 56301
Contact: Helen Saatzer

Mississippi Society of Scientists and Inventors, Inc.
P.O. Box 100
Sandhill, MS 39161

Confederacy of Mississippi Inventors
Rt. 1, Box 244A
Vicksburg, MS 39180
Contact: Rudolph E. Paine

American Society of Inventors, Inc.
402 Cynwyd Dr.
Absecon, NJ 08207
Contact: J. Phil Richey

National Society of Inventors
23 Palisades Ave.
Piscataway, NJ 08854
Contact: James H. Blow, Jr.

Society of American Inventors
P.O. Box 7284
Charlotte, NC 28217
Contact: A. N. Tringali

The Inventors Council of Ohio
P.O. Box 2656
Columbus, OH 43216

Oklahoma Inventors Congress
8300 S.W. 8th St.
Oklahoma City, OK 73108
Contact: Julian Taylor

Appalachian Inventors Group
P.O. Box 388
Oak Ridge, TN 37830
Contact: C. Samuel Hurst

Intermountain Society of Inventors & Designers
P.O. Box 1514
Salt Lake City, UT 84110

Northwest Inventors Association
723 East Highland Drive
Arlington, WA 98223

Government & University Groups

California Polytechnic State University
San Luis Obispo, CA 93407

Connecticut Products Development Corp.
78 Oak St.
Hartford, CT 06106

National Referral Center for Science & Technology
Library of Congress
Washington, DC 20540

The Support Network 199

SCORE (Service Corps of Retired Executives)
U.S. Small Business Administration
1441 L St., N.W., Room 100
Washington, DC 20416
(400 chapters throughout the United States)

NASA Technology Utilization Program
P.O. Box 8756
Baltimore, MD 21240

Massachusetts Technology Development Corp.
131 State St., Suite 620
Boston, MA 02109

Army Materials & Mechanics Research Center
Watertown, MA 02172

PTC Research Foundation
2 White St.
Concord, NH 03301

New England Industrial Resource Development Program
Durham, NH 03824

Office for Promoting Technical Innovation
New Jersey Department of Labor and Industry
Labor & Industry Building
Trenton, NJ 08625

University of New Mexico
Technical Applications Center
Albuquerque, NM 87131

Center for Entrepreneurial Development
Carnegie-Mellon University
Pittsburgh, PA 15213

National Technical Information Center
Springfield, VA 22161

Impact Seven, Inc.
Box 8
Turtle Lake, WI 54889

Also consult the graduate business schools of universities; many offer some aid to inventors.

Other Groups Offering Some Degree of Help

Lockheed Information Systems
3251 Hanover St.
Palo Alto, CA 94304

Invention Marketing and Licensing Agency
121 N. Fir St.
Ventura, CA 91003

Innotech Corp.
2885 Reservoir Ave.
Trumbull, CT 06611

Cambridge Research & Development Group
21 Bridge Square
Westport, CT 06880

Innovator Associates
847 W. Newport Ave.
Chicago, IL 60657

Institute for Invention & Innovation, Inc.
85 Irving St.
Arlington, MA 02174
(This is Richard Onanian's company; he publishes *Invention Management*.)

American Research & Development Corp.
1 Beacon St.
Boston, MA 02108

Control Data Worldtech, Inc.
474 Concordia Ave.
SCNFAC
St. Paul, MN 55103

Gulf & Western Invention Development Corp.
1 Gulf & Western Plaza
New York, NY 10023

National Patent Development Corp.
375 Park Ave.
New York, NY 10022

Product Resources International
90 Park Ave.
New York, NY 10016

Rainhill Group, Inc.
80 Wall St.
New York, NY 10005

Research Corporation
405 Lexington Ave.
New York, NY 10017

Center for New Business Executives
400 Oberlin Rd.
Suite 350
Raleigh, NC 27605

Scientific Advances, Inc.
1375 Perry St.
Columbus, OH 43201

Center for Venture Management
207 East Buffalo St.
Milwaukee, WI 53202

APPENDIX H

RECOMMENDED READING

This list includes books, periodicals, and other publications that you may find useful. General reference works and periodicals appear first, followed by titles organized according to the chapters in this book to which they are most relevant. Many of these can be found in your local library. In some cases ordering information has been given.

General Information and Sources

Air Conditioning, Heating, and Refrigeration News, Directory Issue. P.O. Box 2600, Troy, MI 48084.
   Free with a subscription to the journal or $10 if purchased separately.
Chemical Engineers' Handbook. Robert H. Perry and C. H. Chilton. New York: McGraw-Hill Book Co., 1973.
Electronics Engineer's Handbook. Donald G. Fink and Donald Christiansen. New York: McGraw-Hill Book Co., 1982.
Machine Design magazine, 1111 Chester Avenue, Cleveland, OH 44114.
   23 regular issues plus 5 reference issues, $50.
Marks' Standard Handbook for Mechanical Engineers, 8th ed. Theodore Baumeister. New York: McGraw-Hill Book Co., 1978.
Material Selector. Materials Engineering, P.O. Box 91368, Cleveland, OH 44101.
   Free with free subscription to *Materials Engineering*.
Mechanical Products Catalog. Hutton Publishing Co., 333 Broadway, Jericho, NY 11753. $35.

Modern Plastics Encyclopedia. Modern Plastics, P.O. Box 602, Hightstown, NJ 08520.
   Engineering; design; textbook and directory. Free with a $24 subscription to *Modern Plastics*.
Plastics Directory. Plastics World, Computer Center, P.O. Box 5391, Denver, CO 80217.
   Free with a free subscription to *Plastics World*.
Standard Handbook for Electrical Engineers. Donald G. Fink and H. Wayne Beaty. New York: McGraw-Hill Book Co., 1978.
Thomas Register of American Manufacturers and *Thomas Register Catalog File*. Thomas Publishing Co., 1 Penn Plaza, New York, NY 10001.
   Products and services; trade names; company names and addresses, ratings, officers; company catalogs, cross-indexed. 14 volumes, $85.
Thomas Regional Directory. Thomas Publishing Co., 1 Penn Plaza, New York, NY 10001.
   Same information as above, published on a regional basis. 1 volume, free to the companies listed; otherwise, $35 for each region.
U.S. Industrial Directory. 999 Summer St., P.O. Box 3809, Stamford, CT 06905.
   Company addresses, phone numbers; products; literature and catalogs. 4 volumes, $75.
Used Equipment Directory. 70 Sip Ave., Jersey City, NJ 07306.
   Paperback, $2.

See also:
Texts on all aspects of metals. American Society for Metals, Book Order Department, Metals Park, OH 44073.
Texts on industrial and mechanical engineering. Wiley Professional Books-by-Mail, Dept. 0121, Somerset, NJ 08873.
Texts of scientific and engineering subjects (such as the *Handbook of Chemistry and Physics*). CRC Press, Inc., 2000 Corporate Boulevard, NW, Boca Raton, FL 33431.
Trade magazines and directory issues in many fields. McGraw-Hill Book Co., 1221 Avenue of the Americas, New York, NY 10020. Also available from Penton/IPC, 614 Superior Avenue West, Cleveland, OH 44113.
The Yellow Pages, Business-to-Business volume; Industrial Purchasing Yellow Pages, available in some metropolitan areas.

Journals on Invention and New Products

Action. Association for the Advancement of Invention & Innovation, 1911 Jefferson Davis Hwy., Arlington, VA 22202.

Recommended Reading

American Inventor. 10310 Menhart Lane, Cupertino, CA 95014.
Dialog. Lockheed Information Systems, 3251 Hanover St., Palo Alto, CA 94304.
Idea. PTC Research Foundation. 2 White St., Concord, NH 03301.
Inside Research & Development. 2337 Lemoine Ave., Fort Lee, NJ 07024.
International Invention Register. P.O. Box 547, Fallbrook, CA 92028.
International New Product Newsletter. 390 Stuart St., Boston, MA 02117.
Invention Management. Institute for Invention & Innovation, 85 Irving St., Arlington, MA 02174.
Inventors Digest. American Society of Inventors, 947 Old York Rd., Abington, PA 19001.
The Lightbulb. Inventors Workshop International, 121 N. Fir St., Ventura, CA 91003.
MGA Technology Newsletter. 2 East Oak St., Chicago, IL 60611.
Man Tech Journal. Army Materials & Mechanics Research Center, Watertown, MA 02172.
New Product–New Business Digest. General Electric Company, 1 River Road, Schenectady, NY 12345.
New Products and Processes. Newsweek, 444 Madison Ave., New York, NY 10022.
New Technology Index. 1105 Market St., Wilmington, DE 19801.
Patent Licensing Gazette. 37 Easton Rd., Willow Grove, PA 19090.
R&D Management Digest. P.O. Box 56, Mt. Airy, MD 21771.
Tech Briefs. NASA Technology Utilization Program, P.O. Box 8756, Baltimore, MD 21240.
Technical Survey. 11001 Cedar Ave., Cleveland, OH 44106.
Technology Marketing. General Electric Company, 1 River Rd., Schenectady, NY 12345.
Technology Transfer Times. 167 Corey Rd., Brookline, MA 02146.
Technotec. Control Data Corp., 8100 34th Ave. S., Minneapolis, MN 55440.
Unit. Dr. Dvorkovitz & Associates, P.O. Box 1748, Ormond Beach, FL 32074.

Chapter 1: Protection

Abernathy, David, and Knipe, Wayne. *Ideas, Inventions, and Patents.* Atlanta: Pioneer Press, 1974.
Invention Management. Newsletter published by the Institute for Invention and Innovation, Inc., 85 Irving St., Arlington, MA 02174. Editor and publisher, Richard A. Onanian. ($96/yr.)

Mandell, Irving. *How to Protect and Patent Your Invention.* Dobbs Ferry, N.Y.: Oceana, 1951.

Pressman, David R. *Patent It Yourself! How to Protect, Patent, and Market Your Inventions.* New York: McGraw-Hill, 1979.

Chapter 2: Patenting

American Bar Association. *What Is a Patent?* Circulation Department, 1155 East 60th St., Chicago, IL.

Attorneys and Agents Registered to Practice Before the U.S. Patent and Trademark Office. Washington, D.C.: Superintendent of Documents, U.S. Government Printing Office, 1980. ($8)

General Information Concerning Patents. Washington, D.C.: U.S. Government Printing Office, 1982. (60 cents)

Greer, Thomas J., Jr. *Writing and Understanding U.S. Patent Claims.* Charlottesville, VA: Michie Co./Bobbs-Merrill, 1979.

Gregory, James. *The Patent Book.* New York: A & W, 1979.

Jacobs, Albert L. *Patent and Trademark Forms*, rev. ed., New York: Boardman, 1978.

Jones, Stacy V. *The Inventor's Patent Handbook.* New York: Dial, 1969.

Kursh, Harry. *Inside the U.S. Patent Office.* New York: Norton, 1959. (A history.)

Richardson, R. O. *How to Get Your Own Patent.* New York: Sterling, 1981.

Whitehurst, Bert W. *How to Make U.S. Patent Searches.* Beverly, Me.: Galleon-Whitehurst, 1978.

Chapters 3 and 4: Financing

American Bar Association. *Considerations in Selecting an Invention Promoter.* Circulation Dept., 1155 East 60th St., Chicago, Ill.

Association for the Advancement of Invention and Innovation. *Information for Inventors—Where to Get Information and Assistance.* 19–11 Jefferson Davis Hwy., Arlington, VA 22202.

Conot, Robert. *A Streak of Luck: The Life and Legend of Thomas Alva Edison.* New York: Seaview, 1979.

Hanan, Mack. *Venture Management.* New York: McGraw-Hill, 1976.

Hayes, Rick Stephan. *Business Loans: A Guide to Money Sources and How to Approach Them Successfully.* Inc./CBI, 1980.

Hellmuth, James G. *Find Money: A Businessman's Guide to Sources of Financing.* New York: Boardman, 1979.

INC. 342 Madison Ave., New York, N.Y. 10017. ($18/yr.)

International Entrepreneurs' Association. *How to Get an SBA Loan.* 631 Wilshire Boulevard, Santa Monica, CA 90401.

Josephson, Matthew. *Edison, a Biography.* New York: McGraw-Hill, 1959.

Kelley, Joseph C. *Making Inventions Pay.* New York: McGraw-Hill, 1950.

Martin, Robert B., Jr., and Thompson, Davis D. "Planning for the R&D Tax Shelter: An Analysis of the Essential Tax Elements." *The Journal of Taxation,* October, 1980.

Moore, N. G., et al. "How Limited Partnerships Tax-shelter the R&D of New Products or Technology." *The Journal of Taxation,* September, 1978.

Pratt, Stanley E., ed. *Guide to Venture Capital Sources.* Wellesley Hills, Me.: Capital Publishing Corp., 1981.

Pratt, Stanley E., ed. *How to Raise Venture Capital.* New York: Scribners, 1982.

Rao, D., ed. *Handbook of Business Finance and Capital Sources.* New York: American Management Association, 1980.

Shemin, Paul S. "Consumer Law—Idea Promoter Control: The Time Has Come." *New York Law Journal,* March 30, 1978.

"Tax Consequences of Patents, Trademarks, Trade Secrets, and Copyrights." *American Patent Law Association Journal,* Vol. 3, No. 1, 1975.

White, Richard M. *The Entrepreneur's Manual: Business Start-Ups, Spin-Offs and Innovative Management.* New York: Chilton, 1980.

Venture. 35 West 45th St., New York, NY 10036. ($18/yr.)

Chapter 5: Development On Your Own

"The Invention Development Phenomenon." *American Patent Law Association Journal.* Vol. 6, No. 2, 1978. Suite 203, 2001 Jefferson Davis Highway, Arlington, VA 22202.

Lindberg, Roy A. *Processes and Materials of Manufacture.* (Boston: Allyn & Bacon, 2nd edition, 1977).

Standards and other information available from:

American Society for Testing and Materials, 1916 Race St., Philadelphia, PA 19103.

American Society of Heating, Refrigerating, and Air Conditioning Engineers, 1791 Tullie Circle, NE, Atlanta, GA 30329.

American Society of Mechanical Engineers, 345 East 47th St., New York, NY 10017.

National Bureau of Standards, Washington, D.C. 20234.
National Technical Information Service, 5285 Port Royal Rd., Springfield, VA 22161. (Telephone: 703-487-4640)

Chapter 6: Contracting Out Development and Manufacturing

"Invention Development Services and Inventors: Recent Inroads on Caveat Inventor." *Journal of the Patent Office Society*, Vol. 60, No. 6, June, 1978.
Small Business Administration. *Decision Points in Developing New Products*. Publication No. 39, Washington, D.C.

Chapter 7: Licensing or Sale

American Bar Association. *Sorting Out the Ownership Rights in Intellectual Property: A Guide to Practical Counseling and Legal Representation*. Circulation Department, 1155 East 60th St., Chicago, Ill.
Finnegan, Marcus B. *1979 Licensing Law Handbook*. New York: Boardman, 1980.
Finnegan, Marcus B., and Goldscheider, Robert. *The Law and Business of Licensing*. New York: Boardman, 1975.
Gleeson, Murray A. *How to Make Money from Your Patent Invention*. DuPage Products Co., Box 261, Elmhurst, Ill. 60126.
Klein, Paul, Einhorn, H., and Ancier, W. A. *Patent Licensing Transactions*. New York: Matthew Bender, 1968.
Nierenberg, Gerald I. *The Art of Negotiating*. New York: Hawthorne, 1968.
Reefman, William E. *How to Sell Your Own Invention*. North Hollywood, Calif.: Halls of Ivy Press, 1977.

Chapter 8: Setting Up Your Own Plant

Baumbeck, Clifford M., and Lawyer, Kenneth. *How to Organize and Operate a Small Business*. Englewood Cliffs, N.J.: Prentice-Hall, 1979.
Brummet, Lee, and Anderson, Jack. *Cost Accounting for Small Manufacturers*. An SBA Publication, Small Business Administration, Washington, D.C.

Burstiner, Irving. *The Small Business Handbook: A Complete Guide to Starting and Running Your Own Business.* Englewood Cliffs, N.J.: Prentice-Hall/Spectrum, 1980.

Dible, Donald M. *Up Your Own Organization!* Santa Clara, Calif.: The Entrepreneur Press, 1978.

Genera, Robert L. *The Employer's Guide to Interviewing.* Englewood Cliffs, N.J.: Prentice-Hall, 1979.

Mancuso, Joseph. *How to Start, Finance, and Manage Your Own Business.* Englewood Cliffs, N.J.: Prentice-Hall, 1978.

Smith, Irene. *Diary of a Small Business.* New York: Scribners, 1982.

Stevens, Mark. *How to Run Your Own Business Successfully.* New York: Monarch Press, 1979.

Zwick, Jack. *A Handbook of Small Business Finance.* SBA Management Series No. 15, 1975. Small Business Administration, Washington, D.C.

Chapters 9 & 10: Marketing

Breen, George F. *Do It Yourself Market Research.* New York: McGraw-Hill, 1977.

Culligan, Matthew J. *Getting Back to the Basics of Selling.* New York: Crown, 1979.

Grikscheit, Gary M., Cash, Harold C., and Crissy, W. J. E. *Handbook of Selling.* New York: Wiley, 1980.

"Advertising Small Business." *Small Business Reporter.* Vol. 15, No. 2, Bank of America, Dept. 3120, P.O. Box 37000, San Francisco, CA 94137. (2) 1981

Oathout, John D. *Trademarks: A Guide to the Selection, Administration, and Protection of Trademarks in Modern Business Practice.* New York: Scribners, 1981.

Stern, Louis, and El-Ansary, A. I. *Marketing Channels.* Englewood Cliffs, N.J.: Prentice-Hall, 1977.

Tolly, Stuart. *Advertising and Marketing Research.* Chicago: Nelson-Hall, 1977.

Zoltman, Gerald, and Burger, Phillip. *Marketing Research.* Hinsdale, Ill.: Dryden Press, 1975.

Chapter 11: New Fields to Conquer

Adams, James L. *Conceptual Blockbusting—A Guide to Better Ideas.* San Francisco: Freeman, 1974.

Buzan, Tony. *Use Both Sides of Your Brain.* New York: Dutton, 1974.

Bylinsky, Gene. *The Innovation Millionaires: How They Succeed.* New York: Scribners, 1976.

Calvert, Robert, ed. *The Encyclopedia of Patent Practice and Invention Management.* New York: Van Nostrand Reinhold, 1968.

Dorland, Gilbert, and Van Der Wal, John. *The Business Idea.* New York: Van Nostrand Reinhold, 1976.

Green, Orville N., Durr, Frank L., and Berseth, John. *Practical Inventors Handbook.* New York: McGraw-Hill, 1979.

Johnson, Alfred R. *The How of Invention Management.* Cambridge, Mass.: Arthur D. Little, Inc., 1977.

Kivenson, Gilbert. *The Art and Science of Inventing.* New York: Van Nostrand Reinhold, 1977.

Mamis, Robert A. "Hoyle Schweitzer's Decade of Discontent." *INC.,* February, 1982.

Mueller, Robert E. *Inventivity: How Man Creates in Art and Science.* New York: John Day, 1963.

APPENDIX I

INVENTORS' SHOWS

IWI Inventors Exposition
121 N. Fir St.
Ventura, CA 91003

World Fair for Technology Exchange
Dr. Dvorkovitz & Associates
P.O. Box 1748
Ormond Beach, FL 32074

Mid-American New Ideas Fair
P.O. Box 100
Hill City, KS 67642

National Inventors Show
New York Hilton Hotel
Avenue of the Americas at 53rd St.
New York, NY 10019

Minnesota Inventors Congress
P.O. Box 71
Redwood Falls, MN 56283

Cleveland Engineering Society
3100 Chester Street
Cleveland, OH 44114

Appalachian Inventors Fair
P.O. Box 388
Oak Ridge, TN 37830

Index

abstracts, 25–27
advertising
  cooperative, 160–61
  for investors, 63–64
American Gas Association, 102
American Research and Development Corporation (ARD), 72
American Society for Testing and Materials, 98
approvals, 101–3
architect, product, 105
assembly jigs, 114

background, writing, 9, 27–29
bankruptcy, 143–44
barter, 70
bonus/penalty, 118
Booz-Allen & Hamilton, 133–34
borrowing, 71
Brill, Ray, 165
business
  plan, 64–67
  starting, 69–72

Cairns, Walter J., 173
capital gains, 85
certification, 101–3
Chandler, William R., 61
claims, 44–52
coinventors, 113
company laboratories, 13
computers, 169–72
confidentiality, 109, 119
contractor, general, 105
contracts
  termination, 109–10
  development, 108–12

conventions, 14–15
copyright, 1
corporate finance, 84
corporation, forming, 67–68
cost estimates, 100–101
crafts catalogs, 10
customer payments, collecting, 141

description, of invention, 9, 31–44
design, of invention, 108
designer, industrial, 105–6, 115
developers, 106–7
development, 90–103
diagrams, schematic, 5
diary, 2–4
Dingee, Alexander L. M., 62
direct mail, 156–57
distributors, 149–52
drawings, 25–44
  engineering, 5–6
drop-shipping, 150–51

entrepreneurs, 61–62
equipment, buying, 138–39
European Patent Convention (EPC), 53
evaluating the invention, 13–16
exclusivity, and licensing, 129
exhibits, 14–15

factory overhead, 101
finances, 109
  difficulties, 143–45
foreign filing, 52
foreign patent, 173
fraud, 57–61

General Information Concerning Patents, 52
Goldschieder, Robert, 126–27
government assistance, 68–69
Guide to Venture Capital Sources (Dingee *et al.*), 62, 77, 82

Haslett, Brian, 62
House Science and Technology Committee, 165

improvements of products, 112–13
industrial design, 115
infringement, 54–55
 and licensing, 132
innovation, 16, 165–66
intellectual property, 108
interference, 54
International Licensing Network, Ltd., 126–27
Invention Management, 14, 69, 133–34
investors, finding, 61–64

joint ventures, 173–74

laboratories, 13
laboratory testing, 97–98
lawsuits, 103
layouts, 95
legal advice, 18, 132
licensee, 124–27
licensing, 56–57, 128–32
 legal advice, 132
 termination, 132
 vs. selling, 127–28
Lindberg, Roy A., 94
Little, Arthur D., 69, 173

mail survey, 108
manufacturers, 14–15

manufacturer's representative, 156, 161–63
manufacturing
 directory, 10
 independent, 135–45
marginal design, 99
marketing, 154
 independent, 155–63
 national distribution, 149–52
markets, 14, 15, 163
materials, 94, 100
models, 10–12
 building, 107–8
money management, 141–43

National Bureau of Standards, 68–69, 98
national distributors, 149–52
Nerad, Tony, 97
notarized disclosure, 2

office actions, 53
Onanian, Robert, 14, 69
options, 132–33
Original Equipment Manufacturer (OEM), 100, 116, 146–49
ownership, 116

partnerships, 72–73, 79–84
parts suppliers, 14
Patent Cooperation Treaty (PCT), 52–53
Patent Exchange, 73–75
Patent Office, U.S., 10
 appeals, 55, 168
patent pending, 54
patents, 1, 9
 applications, 19–22
 cost, 18
 design, 10, 22
 examiners, 19
 filing, 19, 52–54
 law, 10

patents (*continued*)
  living-cell, 23
  materials, 23
  on parts, 19
  preparing own, 18
  process/method/system, 23
  protection, 53
  reading, 22
  structure, 23
  writing, 51–52
personnel
  hiring, 93–94, 136–37
  motivating, 137–38
preliminary disclosure, 8–10
pricing, 157–58
prior art search, 23–25
processes, 94
Processes and Materials (Lindberg), 94
product development, 105–8
production, designing for, 94–96
product liability insurance, 99
professional societies, 97–98
prototype, 10–12, 85
public stock issue, 86–89
purposes, of invention, 29–31

quality control, 118

registered letters, 2
research and development, 90
right to buy elsewhere, 152
royalties, 85, 173
  and licensing, 130

sales, direct, 156–57
searches, 23–24
Securities and Exchange Commission (SEC), 86
  Regulation D, 88
seed capital, 61, 71
shares, 74
Shemin, Paul S., 57–59
sketches, 4–8
  perspective, 6–8

Small Business Administration (SBA), 68–69
Small Business Investment Companies, 77
Smollen, Leonard E., 62
specifications, 25–44
staff. *See* personnel
standards, 98
stocks, 71
strategies, 14
subcontracting, 115–23
success rate, 14
suppliers, finding, 139–40

tax shelters, 85–86
testing, 13, 118–19
  field, 98–100
  laboratory, 97–98
  standards, 97
Thomas Register, 10, 92–93, 95, 140, 153–54
tolerances, 96
tools, special, 113–14
trade magazines, 10
trademark, 10, 153–54
trade shows, 14–15
trends, 76

Underwriters' Laboratories, 102
U.S. Court of Customs and Patent Appeals, 54
U.S. Patent Office. *See* Patent Office, U.S.

venture capital, 61–62, 77–79, 87
Venture magazine, 88–89

Wallace, Peter W., 87–88
warranties
  and licensing, 130–31
  provisions, 119
Wetzel, William, 78
wholesale vs. retail, 158–61
witnesses, 2